Plants of the
Oregon Coastal Dunes

Plants of the
Oregon Coastal Dunes

Alfred M. Wiedemann
La Rea J. Dennis
Frank H. Smith

Oregon State University Press
Corvallis

The paper in this book meets the guidelines for permanence and durability of the Committee on Production Guidelines for Book Longevity of the Council on Library Resources and the minimum requirements of the American National Standard for Permanence of Paper for Printed Library Materials Z39.48-1984.

Library of Congress Cataloging-in-Publication Data
Johnston, L.D. (La Rea Dennis)
Plants of the Oregon coastal dunes / La Rea J. Dennis Johnston, Alfred M. Wiedemann, Frank H. Smith.
    p.   cm.
Includes index.
ISBN 0-87071-457-0 (alk. paper)
1. Sand dune plants—Oregon—Identification. 2. Coastal plants—Oregon—Identification. 3. Sand dune plants—Oregon—Pictorial works. 4. Coastal plants—Oregon—Pictorial works. 5. Plant communities—Oregon. I. Wiedemann, Alfred M. II. Smith, Frank H. III. Title.
QK182.J646 1999
581.7'583'09795—dc 21   98-49879
                                                              CIP

Printed in the United States of America

Oregon State University Press
101 Waldo Hall
Corvallis OR 97331-6407
541-737-3166 •fax 541-737-3170
http://osu.orst.edu/dept/press

Contents

Introduction

This book was originally published in 1969 and has, for nearly 30 years, served as a guide to the natural history of the Oregon coastal dunes and to the plants found on them. In this edition we have updated the scientific names of the plant species to correspond with those in current use in local floras.

During these past thirty years changes have occurred in the dune landscape. The deflation plains, developing on the lee side of the foredune and between the lateral ridges of the large parabola dunes, were vegetated mostly by meadow species, with a few tree and shrub seedlings. Today impassable thickets of trees, coast pine (*Pinus contorta*) and Sitka spruce (*Picea sitchensis*) and shrubs western wax myrtle (*Myrica californica*), coast willow (*Salix hookeriana*) and evergreen huckleberry (*Vaccinium ovatum*) have grown up, requiring trails to be cut for access to the beach.

On the foredune and other areas of active, blowing sand, the continued spread of European beach-grass (*Ammophila arenaria*) has resulted in the suppression of the native dune-building and sand-stabilizing species. While not generally threatened with extinction, many of these species, such as the yellow abronia (*Abronia latifolia*), gray beach pea (*Lathyrus littoralis*), beach morning-glory (*Calystegia soldanella*), American dune-grass (*Elymus mollis*) and large-headed sedge (*Carex macrocephala*) are no longer seen in the abundance of thirty years ago. One endangered species, pink sandverbena (*Abronia umbellata* ssp. *breviflora*) which was once abundant along the coast from British Columbia to northern California is now restricted to a few sites from the central Oregon coast south. Since 1991 efforts have been underway by the Plant Conservation Biology Program of the Oregon Department of Agriculture to reintroduce this species to create new populations. Reintroduction has been most successful in sites where European Beach-grass has been brought under control.

Control of European beach-grass has been undertaken, with some success, in certain areas to improve the habitat for an endangered bird, the snowy plover (*Charadrius alexandrinus* ssp.

nivosus). In addition, biological control agents (primarily seed weevils and the Gorse Spider mite) have been introduced by the Oregon Department of Agriculture in cooperation with other agencies and private land owners to help control the spread of gorse (*Ulex europaea*) and Scotch broom (*Cytisus scoparius*).

The main purpose of this book is to help visitors to the sand dunes, regardless of their background, to become a little more familiar with the landscape they see, and with the forces of nature operating within that landscape. A great deal of satisfaction can be obtained through knowing something of the story behind the things one sees in nature: the effect of climate, why plants grow where they do, the names of the plants, etc.

The sand dunes of the Oregon coast offer some of the most spectacular seashore landscapes to be found anywhere in the country. Readily accessible by highway, they have become a source of interest to increasing numbers of travelers to the Northwest. Oregon's excellent State Park system and the establishment of the Oregon Dunes National Recreation Area (ODNRA) also draw large numbers of visitors to the dunes areas.

The first chapter of this book deals with the sand dune area as a whole—its geologic history, climate, dune forms and formation of dunes, and some general observations on the vegetation. Much of this information was put together from many sources, and numbers in parentheses throughout the text refer to a bibliography which lists the major sources of information.

The next four chapters discuss the plants and plant communities of the sand dunes. The information is based largely on the dune areas from Tillamook Head to Coos Bay. Many plants of the Clatsop area dune ridges are included, but this area has many additional species that have been introduced by the activities of man. Likewise, the coast below Coos Bay has a slightly different climate from that of the rest of the Oregon coast, and consequently there are many additional plant species present.

The plants included in this book are only those that might be considered common on the dunes. It does not include at least as many more that are less common. It does not include roadside plants of the "non-dune" coastal forest, except when these occur on the dunes.

Common names are used in the discussion with the scientific name given in parentheses the first time it is used in a chapter. Whenever possible, it is desirable to learn the scientific names of plants, since common names can be a source of much confusion.

The taxonomic key of Chapter 6 includes ninety of the plants one is most likely to see on the sand dunes. Instructions for its use precede the actual key. In addition to the plants in the key, forty other species are described in the description section. Identifying characteristics are based on flowers and leaves, which would make it most useful during the spring and summer months. However, the key, together with the photographs and descriptions of the plants in Chapter 7, should make it possible to identify many of these species throughout the year. Mosses and lichens are not included in the key. A glossary is included that gives definitions of commonly used botanical terms.

A number of individuals assisted in the preparation of this work; namely, Kenton L. Chambers, William W. Chilcote, Ronald J. Tyrl, the late Weldon K. Johnston, Richard R. Halse and Thomas N. Kaye. We wish to thank them for their valuable assistance.

The photographs were taken by the late Frank H. Smith, with the exception of *Ceanothus thyrsiflorus* (blue blossom), which was kindly furnished to us by Mr. and Mrs. Orin Hess of Wedderburn, Oregon.

The Oregon Coastal Dunes

Sand dunes are found along the entire west coast of North America from Alaska to Lower California. They attain their greatest development along the coast of Oregon and the southern coast of Washington. Of some 310 miles of ocean-facing shoreline in Oregon, there are dunes on 140 miles or 45 percent of the total. The Washington dunes, extending 55 miles north of the mouth of the Columbia River, can be considered a part of the Oregon dunes system.

Several prominent physiographic features characterize the Oregon coastline. Most spectacular are the high, steep sea cliffs and promontories composed of erosion-resistant igneous rock. These alternate with low, narrow coastal plains on which the dune systems develop. Both are joined to the Coast Range mountains at their inner margins. All the streams and rivers flowing into the ocean show evidence of "drowning"—a result of a relatively recent rise in level of the ocean. The former streambeds of some of the larger rivers, such as the Columbia and Coos, can be traced across the continental shelf for a considerable distance.

Although all of the present dunes are more or less active, there are a number of locations where very old, semi-consolidated dunes have been exposed (mostly through construction activities). Strikingly different in color and soil development, they suggest a very ancient development of dune systems.

The sand dune areas along the coast can be readily subdivided into four regions that were first described by Cooper.[1] These subdivisions are based both on shoreline features and on the forms of the dunes themselves, and corresponds quite closely to various shoreline processes to be outlined later.

The first region is that associated with the mouth of the Columbia River and is a continuous belt of dunes extending 55 miles north in Washington and 18 miles south in Oregon to Tillamook Head. It is the result of deposition of erosion sediments from the Columbia River which have created a prograding (advancing) shoreline. Characteristic mainly of this dune area are the series of sand ridges running parallel to the shore. This is the only location where the parallel ridge system occurs on the Oregon coast, although it is common on the east coast of the United States and in Europe.

The second region extends from Tillamook Head south some 125 miles to Heceta Head. There are many capes and headlands along the north and central coast so that the dune areas tend to be isolated and quite variable in size and form. Of fourteen locations with significant dunes, eight are associated with bays or estuaries. Practically all are characterized by a large parabola dune or parabola dune complex (see page 17) which has developed as a result of peculiar vegetation-wind interactions. Marine erosion has left these parabola dunes in various stages of degradation (erosion of their seaward ends) from nearly complete systems to mere remnants still remaining.

The most extensive and impressive dunes are in the third region, the coast between Heceta Head and Coos Bay, a distance of 54 miles. The dunes are continuous except where two major rivers, the Siuslaw and Umpqua, and several smaller streams cross them. They rest on the broad, low surface of a rock terrace which slopes gently below sea level and extends inland up to two and one half miles. Cooper calls this area the "Coos Bay dune sheet" and attributes its significance to the fact that "the great extent and continuity of receptive shore backed by terrain favorable to dune migration give ample opportunity for development of materials and forces."[2]

The fourth region is comprised of the dunes between Cape Arago and the California border. The dunes of this region are low and flat and extend inland for a considerable distance; the most notable extending north and south of the Coquille River for a total distance of about 12 miles. They are very low and extend inland about a mile. Smaller areas occur at various locations south to California.

Cooper[1] described in some detail thirty dune locations along the coast of Washington and Oregon (subdividing the Coos Bay dune sheet into five subregions for convenience). Through comparative ground and aerial observations he was able to propose a developmental history more or less common to most of these dune locations. Figure 1 (page 7) is an adaptation from Cooper's Plate 1 showing the distribution and location of the sand dune areas and physiographic features along the Oregon coast. Topography, geologic processes, climate and vegetation have all combined to create the existing landscape, which has a diverse and complex developmental history.

Geology

The geologic history of the Oregon coast is not yet thoroughly understood, and even a general account must rely on several sources.[3] The entire coast falls into two rather distinct divisions separated by the Coquille River. To the south are the Pre-Tertiary (prior to sixty million years ago) formations of metamorphic rock, and to the north are Tertiary and Pleistocene sedimentary depositions and igneous intrusions.

The sedimentary formations had their origin in the early Eocene (sixty million years ago) when a geosyncline (a downward flexure of the earth's crust) occupied the area from the Klamath Mountains north to Vancouver Island and east to the Cascade Mountains, the entire region being under water. The initial deposits were volcanic in origin, but by middle Eocene, uplift activity to the south and subsequent erosion contributed arkosic sands (sands derived from the rapid decomposition of granite and containing feldspar) as marine deposits for a considerable distance northward. Volcanic activity persisted throughout the Eocene so that igneous materials are frequently interbedded with the sedimentary layers. By late Eocene tuffaceous silts (sediments of volcanic origin) and clays, rich in organic matter, began to be deposited by the streams and rivers flowing from the surrounding highlands.

This deposition continued and increased considerably during the Oligocene (beginning forty million years ago), when arkosic sands and silts were also laid down in great amounts. Cape Kiwanda on the north coast, near Pacific City, is an example of Miocene sandstone and siltstone resting on Oligocene mudstone.

During middle Miocene (twenty million years), vigorous volcanic activity formed many of the basaltic intrusions and headlands which remain today as erosional remnants (Yaquina Point, Cape Lookout, the numerous "seastacks," Saddle Mountain). Toward the end of the Miocene, uplift began which was to form the Coast mountains. This uplift reached its

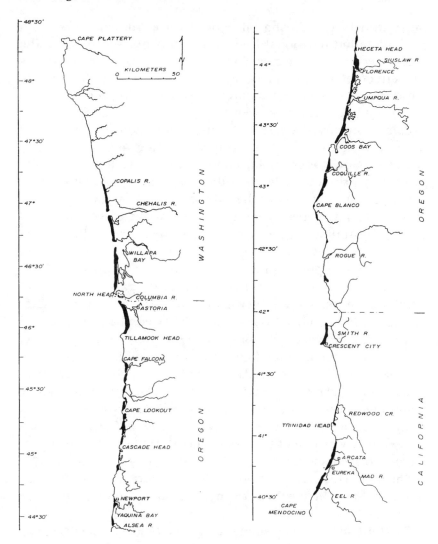

Figure 1. The sand dune localities and major land features of the Pacific Northwest Coastal Region.

maximum height during Pliocene (twelve million years), and peneplanation (erosional leveling of land masses) began. The record becomes a little more complex here, since the coast shows the effect of a number of changes in sea level which are superimposed upon the apparently slowly rising coastline.

During late Pliocene and early Pleistocene (one million years ago) there occurred a period of extreme submergence, indicated by wavecut terraces found as high as 1500 feet above the present sea level. Subsequent uplift lowered the shore line about 300 feet below present sea level, and it was during this period that the rivers and streams cut their trenches across the continental shelf. Resubmergence to about 160 feet above present sea level saw the beginning of Pleistocene sand dune activity. The location of these "ancient eolian sediments," as Cooper called them, corresponds roughly to present day activity, but appears to have been much more extensive. For the most part destroyed at maximum submergence, they are found today only at the higher elevations not reached by sea action. They can be distinguished by their red and yellow staining, frequent cross-bedding, coarse, well-worn grains and semi-consolidated nature.

A period of relative stability followed by a slight emergence in late Pleistocene resulted in the formation of a terrace that today is situated at an average distance of 150 feet above sea level, though varying from below sea level to some 210 feet above sea level due to warping of the earth's crust. This is the "30-meter terrace" of Cooper, and it is overlain with unconsolidated marine sediments. This terrace is quite distinct along the south coast. It is less so in the north, but some evidence can be found of its existence.

Eustatic activity (elevation changes) since late Pleistocene is associated with the cycles of glaciation; and the last major lowering of the shore line—to 450 feet below present sea level— coincides with the maximum of the last, or Wisconsin, glaciation. According to a discussion by Hansen this was about 20,000 years ago in the west.[4] Another geologist, Fairbridge, places the Wisconsin maximum at 17,000 years ago, and the lowering of shore line at 330 feet.[5] As the glaciers melted, submergence again took place, creating the general features of the present coastline. With resubmergence, sand dune activity began again, the sand moving inland ahead of the advancing sea, reaching its maximum development at the end of the period of submergence, about 6,000 years ago according to both Cooper and Fairbridge.

As a result of this submergence, the coastal rivers and streams are characterized by drowned mouth and valleys. On the Columbia, this drowning (indicated by the upper limits of tide action) extends 140 miles upstream; on the Umpqua, 25 miles. Where rivers have kept pace with the "drowning" by depositing alluvial materials, extensive fertile plains have resulted. The Tillamook area is an example of this. Bays and estuaries result where deposition has not been sufficient, such as Alsea Bay, Yaquina Bay, and Sand Lake.

A special situation exists where small streams not only could not fill their drowned valleys with sediments, but also were at least partiatly blocked by sand dunes moving inland. Large, fresh water lakes were thus formed with their surfaces above sea level. Siltcoos, Tahkenitch, Clear and Eel lakes are among those formed in this manner.

In the period since maximum submergence, the sand dunes have undergone a varying cycle of stabilization and rejuvenation, depending upon vegetation, disturbance, and shoreline processes. During these approximately six thousand years, the shoreline has been in a state of relative stability, though not without activity. Along the north coast, on both sides of the mouth of the Columbia River, there is a prograding shoreline, a result of the tremendous loads of sediments carried by the river. Slight to severe erosion characterizes the central Oregon coastline. The beach between Heceta Head and Coos Bay is the most stable part of the coast, with no progradation or erosion.[6]

Climate

Climate plays no small part in the development and appearance of dune areas, both with regard to its effect on the vegetation, which in turn is very important in dune processes, and through the basic factors of temperature, precipitation and wind. The various systems of climate classification place the Oregon coast in a general category of mild temperatures and abundant precipitation with little or no seasonal deficiency of moisture.

Precipitation and temperature data are shown in Table 1 for two stations, Astoria in the north and North Bend in the south. Latitudinal variations of temperature and precipitation are relatively slight along the entire coast. Annual mean temperature is particularly uniform, ranging only between 51°F and 52°F (between the most northerly and most southerly stations, respectively). Annual precipitation varies somewhat more, with slightly greater amounts occurring toward the north end of the coast.

Fluctuation of average temperature between the warmest and coolest months is also relatively small, ranging between 14°F. difference at North Bend and 21°F. at Astoria. Precipitation is very different, however, showing a very marked seasonal fluctuation. The percentage of the total annual precipitation failing during the summer months of June to September is low, varying from 7 percent at Astoria to 4 percent at North Bend. However, because of low summer evaporation rates (due to cool temperatures) this does not in most cases bring about a moisture deficiency for the plants of this area.

Humidity, cloud, and fog data are very incomplete, but Cooper indicates that these factors contribute significantly to the overall climatic effect of the region.[7] There are some published data on average daily solar radiation and average percentage of possible sunshine for each month of the year for several locations in Oregon.[8] This information shows a considerable increase in solar radiation for the southern third of the coast (as compared to the northern two-thirds) for the low precipitation months of June to

Table 1

	North (Astoria)	South (North Bend)
Temperature (°F)		
Annual Average	51	52
July Average	61	58
January Average	40	44
Precipitation (inches)		
Annual Average	77	65
June to August	5	3
December to February	34	30

August. Under these conditions, the low summer precipitation could have a definite effect on the vegetation.

Wind plays an important part both in its influence on the climate and as an active agent in dune processes. Actual measurements of directional frequency and velocity are scarce. On the basis of nearly twelve thousand observations made at three locations (North Head, Newport, and North Bend) by the U.S. Coast Guard between 1937 and 1942, Cooper summarizes the seasonal wind regime as follows:

> *In summer, onshore winds greatly predominate, and most of these are confined to the sector N.-NW; they cover about one octant seaward from the trend of the coast. Winds within this sector have the greatest average velocity. Fall conditions are transitional; winds of the winter type appear in alternation with those of the summer type. In winter, offshore winds are most frequent but have low velocity, except for the gorge winds of the Columbia River. Onshore winds from south to southwest are relatively infrequent but have by far the greatest velocity; they are parallel to the coast or strike it at an acute angle. Spring conditions are again transitional: N.-NW. winds reappear, alternating with the gentle offshore breezes and high-velocity S.-SW. winds of winter.[9]*

As will be shown shortly, these seasonal wind patterns are very influential in the shaping of the dune landscape.

The Dune Landscape

Sand supply, shore topography, the climatic regime and vegetation are the four interrelated factors involved in determining the location and features of the sand dunes. The locations have been noted. There are a number of topographic features which constitute the basis of the various dune systems. These are the foredune, the parallel-ridge system, the precipitation ridge, the deflation plain, the transverse ridge, the oblique ridge, and the parabola dune. Each will be described briefly. Figure 2 is a diagram showing the relationships of most of these features.

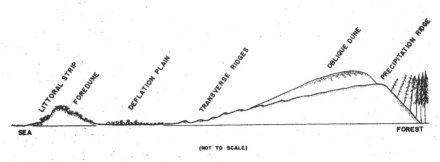

Figure 2.

Sand Supply

The ocean is the great reservoir of sand that has built and maintained the present dune areas. The ultimate sources of these sands are the sediment loads of the rivers and streams, and the coastal Tertiary and Pleistocene formations undergoing marine erosion. Offshore ocean currents distribute this sand, depositing it in places and under conditions not yet clearly understood. The offshore currents flow northward in winter and southward in summer under the influence of the annual wind patterns. Mouths of streams and headlands provide obstacles to this flow such that some of the sand load is deposited and moved up onto the adjoining shore by wave action. The localized dune systems of the central coast, the spits, and the sand fills behind jetties all find their origin in these processes.

One geologist, Twenhofel, theorizes that the bulk of the present dunes are composed of material from more southerly points deposited prior to the fast maximum submergence.[10] This appears to be based on the observation that the black minerals found in the sand have their source mostly south of the Coquille River. Certainly some period of accelerated sand deposition must have occurred to account for the vast amounts of sand now present.

The present-day coastal rivers probably contribute considerably to the sand supply. Particularly outstanding is the Columbia River, which has a water discharge of 7,200 cubic yards per second. By contrast, the Umpqua, the largest river between the Columbia and Cape Blanco, discharges only 265 cubic yards of water per second. Much more sand is delivered by the Columbia than the offshore currents can possibly carry away, resulting in the very extensive parallel-ridge system and prograding shore of this area. The same is also probably true of the Siuslaw and Umpqua rivers, and may account for some of the sand in the massive dunes of this area.

The Foredune

A comparatively recent phenomenon along the Oregon coast is the foredune, the high (up to 25 feet) ridge of sand paralleling the shore immediately above high tide line. The foredune appears to have developed mostly since the 1930s. It is strictly a product of vegetation, and of one plant in particular: European beach-grass *(Ammophila arenaria)*. This grass, introduced in the late 1800s from Europe where it has been used for centuries in dune control work, has spread and been planted all along the Oregon coast. It attains maximum growth and vigor where sand deposition by wind is greatest—the immediate shore area—and the effect is to reduce drastically the amount of sand moving inland off the beach.

The ultimate height and breadth of this type of dune is not known, nor has it been investigated in this country as far as is known. Studies have been conducted in England concerning the formation and possible fate of the foredune; and several early German books give detailed instructions for its construction and maintenance.[11]

The Parallel-Ridge System

The parallel-ridge system is essentially a foredune system, but formed on a prograding shore. Although native plants can initiate such a system, they do not control it as does *Ammophila*. Once started, a large supply of sand builds the ridge rapidly. As the shore extends seaward, the wind deposits sand along a new line of vegetation at the high tide level, less and less being carried inland to the older ridge. Over a long period of time, a considerable number of such ridges will develop (nine on the

Large oblique dune and smaller transverse dune north of Florence near Lily Lake, with deflation plain in foreground.

coast at the mouth of the Columbia River). They are generally 15 to 40 feet high, though they reach 75 feet in places. Figure 3 is a diagrammatic representation of a parallel-ridge system based on the dunes around the mouth of the Columbia River.

The Precipitation Ridge

In the absence of the foredune, or where there is a sufficient sand supply behind this dune, sand is moved inland by the driving force of the seasonal winds. If an area of vegetation is encountered, sand begins to accumulate as a ridge in front of the vegetation barrier because the wind is deflected upward, loses velocity, and drops its load of sand. This ridge will grow in

Figure 3

height, depending upon the height of the vegetation. The lee side becomes very steep, and after a certain angle is reached (about 33° according to many investigators), sand slips down this face, effectively initiating forward movement of the dune and invading the vegetation, usually a forest. This is the precipitation ridge.

The Deflation Plain

The windward slope of the precipitation ridge is very gradual, and as sand removal continues, the moist sand near the water table is eventually reached. At this point, effective sand movement ceases. This a deflation surface, or, where the area is large, a deflation plain. Such areas are found inland from the foredune, especially on the Coos Bay dune sheet, and in the wake of large, moving dunes. Because of the absence of sand movement and the presence of large amounts of water, the deflation plain becomes a favorable habitat for the initiation of vegetation. An excellent example of such an area is found straight west from Cleawox Lake, south of Florence.

Cross bedding in dune resulting from interaction of summer and winter winds.

The Transverse Ridges

Where an extensive area with abundant sand supply and no vegetation occurs, a transverse ridge pattern develops in which the dune crests are oriented generally at right angles to the northwest winds of summer. The ridges average about 6 feet in height, with slopes steep (33°) on the lee side and gentle (3° to 12°) on the windward side. Distances from one ridge to the next vary from 75 to 150 feet and crest length is highly variable. The pattern is partly destroyed in winter and develops again the following summer. Cooper studied the formative aspects of this type of dune very thoroughly and discussed the requirements for their origin and maintenance: unidirectional airflow, dry surface sand, sufficient sand depth, and a supply of new materials. He also studied wind current behavior using smoke and motion pictures.[12] This feature reaches its best development during the summer months on the broad expanses of sand between Florence and Coos Bay.

The Oblique Ridge

The oblique ridges, named and studied by Cooper, are unique in that they seem to have no counterparts in other dune areas of the world, occurring only on the Coos Bay dune sheet of the Oregon coast.[13] They are so designated because their crests are oriented obliquely to both the northwest and southwest winds. The crest behaves as a transverse ridge with a slipface developing on alternating sides as the wind regime goes through its cycle. These dunes occur in a parallel series, averaging some 550 feet from crest to crest. They rise as high as 165 feet above their base, which often is near the water table and may have developing vegetation. Average length is 3,600 feet with extremes up to about 5,000 feet. At their high, inner ends, they are joined by a precipitation ridge which is actively invading what is left of the original dune forest. Their outer, or seaward, ends taper down to an area of transverse ridges, which in turn usually joins the vegetated deflation plain just inside the foredune.

According to Cooper, these dunes are long lived, and move slowly inland. Although the method of formation of these ridges is unclear, he thinks that the existing systems of oblique ridges originated in precipitation ridges close to the shore and that

Unusually broad expanse of fore dunes north of Florence.

extra-large masses of sand around vegetative barriers served as nuclei for the development of oblique ridges. The mass develops into a high dune as the lower precipitation ridge continues to move inland.

The Parabola Dune

This type of dune can develop only on a large sand mass that is stabilized by vegetation. A break occurs in the vegetative cover, a blowout develops, and if there is a differential resistance of the vegetation to sand movement such that it can progress more rapidly in one particular place, a mass of sand, variable in height according to the vegetation it is overwhelming, begins to move inland parallel to a unidirectional wind force. Lateral widening of the break occurs also, creating a ridge somewhat similar in shape to a parabolic curve. The area between the lateral arms may blow down to water table, creating a deflation plain.

Almost all the dunes between Tillamook Head and Heceta Head (Region II) are parabola systems developed by the southwest winds. There are indications that many of these have been eroded at their seaward ends by marine erosion, only remnants being left. The parabola system at Sand Lake is large and well developed. It is shown in its entirety in Figure 4, a vertical aerial photograph. South of Heceta Head, a series of very high parabola dunes occur between the Siltcoos and Umpqua rivers near Tahkenitch Creek. Three of these are formed by northwest winds and the fourth by southwest winds.

Sand Activity

Attempts to reconstruct the history of sand activity and vegetation development and destruction tend to be highly speculative. By comparing dune ridge patterns as revealed in aerial photographs, Cooper developed his concept of cyclic episodes.[14] Episode I is generally a stabilized precipitation ridge and its windward deflation plain at the inner margins of the present dune areas. Episode II is a secondary precipitation ridge which has moved inland, in some cases stopping short of the first; in others, completely overwhelming and covering it. The precipitation ridge of Episode III is currently active, with vegetation stabilization beginning on the windward deflation

Small transverse dune of summer moving from right to left (northwest to southeast) over surface of dune exposed by winter winds.

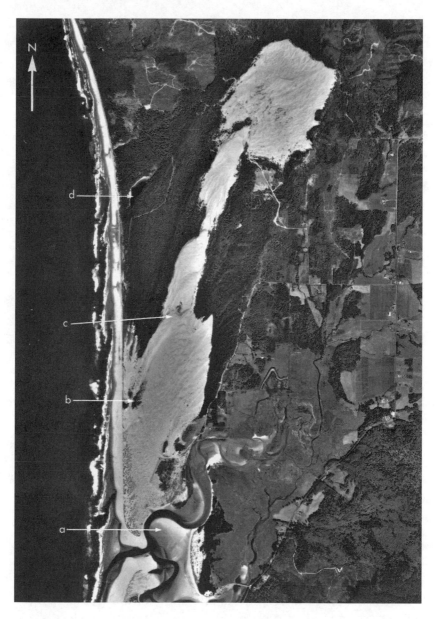

Figure 4. Parabola system at Sand Lake: a. Sand Lake; b. deflation plain; c. parabula dune; d. Chamberlain Lake.

plains. A corresponding cyclic pattern is noted for the parabola dunes. These episodes are difficult to date, but presumably Episode I represents the greatest extent of sand activity, occurring as maximum submergence was reached about six thousand years ago.

Studies by H.P. Hansen on postglacial forest succession corroborate the idea of continual sand activity.[15] He feels that there has been little permanent stabilization of the sand dunes along the ocean shore in the past four to seven thousand years, this being indicated by the presence of sand particles found throughout peat bog borings from the sand dune areas.

Since coast pine *(Pinus contorta)* is the pioneer forest tree on the sand dunes, both on the ridges and in the wet sites and bogs, the proportion of pollen of this species in the peat bog borings can tell something of sand movement. A predominance of pine indicating accelerated movement of sand and destruction of the long established forest. Climate along the coast has apparently had little effect on vegetation development. Pollen samples from peat borings indicate a relatively constant postglacial climate with no warm, dry periods as apparently occurred farther inland. The trees of the climax forest, Sitka spruce *(Picea sitchensis)* and western hemlock *(Tsuga heterophylla)* were never replaced by species characteristic of drier environments.

Along with sand movement, the vegetation has been greatly affected by fire. Fire is one of the causes of renewed sand activity, a fact frequently made evident when old stabilized forest surfaces are exposed by eroding sand. Usually they have charcoal and burned bits of wood in their upper layers and on the surface. Peat bog borings have been found in which charred peat layers occur in conjunction with maximum representation of lodgepole pine pollen.

Cooper discusses fire history in the area around Florence where coast pine—a tree that is one of the first to come up after a fire—grows in almost pure stands.[16] He sees these forests as resulting from two fires in about 1833 and 1853. The Indians around Florence recalled to early settlers that at the time of one of the fires, "The sun was dark for ten days, and nearly all this part of the coast area was burned."[17] The pioneers themselves described "the big dead trees, an endless horde," the result of fire many years previously.[18] In his early descriptions of the Oregon coast, George Davidson, a naval officer, repeatedly emphasized the burned over appearance of the coastal country, concluding: "the whole country has been burned over . . . and bristles with enormous standing trunks of whitened trees."[19]

According to S. N. Dicken, fires caused by humans were widespread during the initial years of settlement from 1845 to 1850.[20] Since then, however, fire has probably been less important as an instrument of change than newer human activities. Logging, grazing, cultivation, mining and the construction of jetties and breakwaters all have helped to alter erosion and deposition patterns along the shore. In his study, Dicken relates these various disturbance factors to physical changes of the shore in the past one hundred years of white settlement.

Vegetation History

The present vegetation of the sand dunes, then, is a result of recurrent sand movement, fires, and most importantly, the influence of man. It is difficult to reconstruct the native vegetation landscape of pre-settlement days, except as the obviously introduced species are eliminated. Early pioneers, who used the beaches extensively for travel, made practically no mention of the vegetation, except, perhaps, as it related to the progress of their journey.

Changes in dune vegetation were traced by B. Hanneson through historical studies.[21] These changes were of two types: (1) changes in species composition through the introduction of plants from other parts of the country or world; and (2) changes in the total area occupied by the various types of vegetation. Working mostly around Bandon in the south, and the Clatsop area in the north, he concluded that most changes were due to disturbance and subsequent stabilization.

Early explorers described the Clatsop spit area south of the Columbia River, and gave some clues as to the native vegetation present. This original vegetation, mostly of prairie types such as bent grasses (*Agrostis*), fescues (*Festuca*), sedges (*Carex*), and clovers (*Trifolium*); and some trees, hemlock (*Tsuga*) and spruce

(Picea), was practically destroyed by sand activity caused by grazing and cultivation. The present vegetation is the result of stabilization work that began in 1935.

The spread of beach-grass *(Ammophila)* is, of course, a dramatic example of change through introduction of new species. Equally impressive is the spread of gorse *(Ulex europaea)* since its introduction as an ornamental in the late 1800s. Covering more than 25,000 acres in Curry, Coos, and western Lane counties, it has resisted most efforts aimed at its eradication.

None of the tree species originally introduced for sand stabilization work have become securely established, and so they have not changed the composition of the original forests. The entire coastal area is included as part of the Cedar-Hemlock Forest by Weaver and Clements in which western hemlock, western red cedar *(Thuja plicata)*, and Sitka spruce are the characteristic tree species.[22] Shelford included the dunes of the Oregon coast in his "Hemlock-Red Cedar-Wapiti Association" of the "Rainy Western Hemlock Forest Biome."[23]

H. D. House published the earliest accounts of the vegetation of the Oregon coastal dunes in 1914.[24] He noted the extensive forests of coast pine with a thick undergrowth of salal *(Gaultheria shallon)*, evergreen huckleberry *(Vaccinium ovatum)*, and western rhododendron *(Rhododendron macrophyllum)*. A discussion of environmental factors and the vegetation is included in N.L.Byrd's study of vegetation zones on the dunes at Waldport.[25] These dunes are presently all in private ownership, and are slowly being subdivided and built into communities and resorts.

A general discussion of climate, dune forms, and dune communities is included by J. J. Reardon in his study of some mammals of the wooded dunes of the middle Oregon coast.[26] His main study on the deer mouse *(Peromyscus maniculatus rubidus)* was conducted on a forested "sand island" directly west of Cleawox Lake. It is thought to be a remnant of the forest standing on these dunes prior to the present period of activity. The forest at present is a mixed stand of conifers: coast pine, western hemlock, Sitka spruce, and Douglas-fir *(Pseudotsuga menziesii)*, with the usual dense understory of salal and evergreen huckleberry.

Plant Communities and Plant Succession

In looking at the landscape, one cannot help but notice that plants are frequently found growing in more or less definite groupings. These groupings are called plant communities. A plant community may consist of only a single species of plant, or it may consist of dozens of species. Likewise its size may be highly variable: it might be a Douglas-fir (*Pseudotsuga menziesii*) forest on a Coast Range mountainside, or it might be the mosses and lichens growing on the top of a rotting wooden fence post. A plant community may be described first, by the plant species which are found in it, and secondly, by the conditions of the environment which directly or indirectly influence which species will become established.

Again, when one thinks of climate or environment the perspective must be kept flexible. There is a climate of the Pacific Northwest and of the Oregon coast. There is also a climate of the north side of a hill which differs from that on the south. Concentrating even more there is a definite climate under the canopy of a forest, another a few inches above the ground, and still another right at the ground surface.

Because each plant species has different growth requirements for such environmental (or climatic) factors as moisture, temperature, and sunlight, a plant community is seldom a stable, non-changing collection of species. Consider what happens to the ground level vegetation when the trees of a forest are removed; or to the vegetation that develops on the fields of an abandoned farm; or on areas of bare ground such as sand or rock.

In every case, certain environmental influences have been modified or changed. On the former forested areas there is bright sunlight instead of shade. Plants that require shade for best growth will slowly lose out and be replaced by those that thrive in full sunlight. Where agricultural fields are no longer cultivated, "weedy" species that depend on human influence no longer can compete against the natural vegetation of the area and they slowly disappear. On bare ground such as sand or gravel, there is no competition from other plants, but the physical environment is often severe and there may be a marked lack of the water and mineral nutrients required by the plant for growth. Here only a very specific kind of plant can become established—one that can grow under these severe conditions.

Each type of vegetative cover literally creates its own special environment in and under it. This special environment may be unfavorable to the plants that created it. In the Douglas-fir forest, for instance seedlings of Douglas-fir cannot grow in the shade of the larger trees. They are "intolerant" of shade. This means that the forest cannot reproduce itself. Seedlings of other trees can, however, grow in the shade of the larger Douglas-firs. Their seedlings will eventually grow into trees which will gradually replace the Douglas-firs. When this replacement is complete, the new trees will be able to continuously reproduce themselves because of the shade tolerance of their seedlings. Such a self-maintaining plant community is called *climax community*; and the sequence of events leading up to it, the continual change of vegetation because of small-scale climatic modifications, is called *plant succession.*

This phenomenon is not limited to forest vegetation. The lichens on a bare rock represent the very first stages of succession by their growth. These small plants slowly decompose the rock surface, and by their death, they add organic material to the rock fragments and so bring about the first stages in soil formation. This little bit of soil allows other plants, perhaps mosses, to get started, and they continue the process. Eventually larger plants become established, and, over a very long period of time, a forest may become established.

Perhaps nowhere can plant communities, and the changes in plant communities—plant succession—be seen as well and as readily as on the sand dunes of the Oregon coast. Because of the continuous movement of these dunes, all types of plant communities, from "pioneer species" on bare sand, to mature forests can be found, often within a short distance. This diversity of plant communities represents visible evidence of plant succession. Because of the mild temperatures—both in summer and winter—and sufficient rainfall, plant establishment and growth is rapid, and marked changes can take place on an area in the relatively short period (in nature's time scale) of a human lifetime.

Sand movement and climate create two general types of habitats (places where plants grow) on the dunes. One is that of the high dune ridges and areas of blowing sand. Here, if the surface is bare of plants, sand blows into dunes which move under the influence of the summer and winter winds. The water table stands far below the surface. Pioneer plant species which become established on such areas must be able to survive extensive sand burial. If they do, they initiate the first stage of a plant successional pattern that will eventually see the sand completely covered and stabilized by forest vegetation.

An almost completely different type of habitat is that of the low, moist deflation plains and other depressions where the sand surface is at or near the water table. Here we find pioneer plants that are characteristic not so much of the blowing sand dunes as they are of places with an oversupply of water. Plant succession is begun by pioneer species that tolerate large amounts of water in their environment, and leads again to a complete covering of the sand surface, eventually by a forest.

Before going into a description of the sand dune plant communities, two things should be made clear. First, although the two general types of habitats—the dry, moving sand dunes, and the wet, stabilized deflation plains—are easy to see and recognize, there are many areas which show various combinations of these. That is, there are transitions from one to the other such that plants of both habitats may be found growing in the same place. This is at least partly due to the second important fact. The sand dune environment can be called a "dynamic system"—something is always happening—a new dune is being formed or an old dune is being stabilized by vegetation. Here a deflation plain is being created by an advancing parabola dune; there, the same dune may be covering an established forest of an earlier deflation plain. A clump of grass builds up a sand mound on a low wet area; on a higher, stabilized dune the disruption of the plant cover may cause the erosion or blowing of sand and the creation of a wet depression. Nowhere is there apparent order and neatness in the sand dunes. Only as one looks closely at the plants and the environments in which they grow will the order of nature be apparent—and from this, hopefully, will come a greater understanding of, and appreciation for, the landscape of which all humankind is a part.

Plant Communities
of the Sand Dunes

Since almost all types of plant seeds will germinate readily in wet sand it is not uncommon to find a wide variety of seedlings on the dunes in the spring as the new growing season begins. Very few of the seedlings survive for long, however, because they have neither the root system to reach down into the deep, moist sand for water, nor can they tolerate the sand burial that is sure to come later.

Only a few plants can do well in this type of environment—plants that could be considered characteristic of sand dune areas. One of the most obvious of these on the Oregon coast is European beach-grass *(Ammophila arenaria)*. Introduced into California in 1896, this grass, a native of Europe, has spread widely on sand dunes the length of the coast both by natural means and through its use in sand stabilization plantings. Mention already has been made of its role in foredune formation, and more will be said of its use in dune stabilization plantings. Where it spreads naturally, through a very vigorous system of creeping underground stems, it creates a very hummocky surface. Individual plants become established and a mound forms as sand accumulates in and around the vigorously growing plant. Such hummocks are frequently found on and near deflation plains. Excellent examples can be found at the end of the gravel road leading west to the dunes just south of Lily Lake, about 8 miles north of Florence. A similar area is located on both sides of the extension of Gallaway Road leading to the beach at Sand Lake, southwest of Tillamook.

Two other plants form large, conspicuous mounds in blowing sand areas. One of these, yellow abronia *(Abronia latifolia)*, grows very close to the sand surface, but has very striking clusters of bright yellow flowers. It arises from a very thick (several inches in diameter) fleshy taproot that penetrates many feet into the sand. Occasionally, sand will be eroded away from these mounds and the thick fleshy root is exposed. The other mound builder is the low-growing silver beach-weed *(Ambrosia chamissonis)*. Its

PLANTS OF THE OREGON COASTAL DUNES

flowers are inconspicuous but the leaves have a thick covering of silvery hairs.

Although these mound-building species are pioneers on open sand, they usually do not start succession toward a complete vegetation cover. Such mounds or hummocks are in a continual state of being built up or eroded away and such areas are very unstable.

The development of a stable plant cover is the function of species which can tolerate sand deposition but which eventually spread over large areas and cover the sand to protect it from movement by wind. Such plants show many characteristics which enable them to survive the hardships of their environment. Seashore bluegrass *(Poa macrantha)* and large-headed sedge *(Carex macrocephala)* both have large seeds which are easily caught up and spread by the wind. These plants also have a very vigorous system of underground stems, and in some places can be seen to literally "invade" bare sand areas.

Gray beach pea *(Lathyrus littoralis)* and beach silver-top *(Glehnia littoralis)* both have large seeds, and have stems which can come up through many feet of sand. The seedlings in particular can send long roots deep into the sand to provide a good supply of water for the developing plants. The beach morning-glory *(Calystegia soldanella)* spreads rapidly by means of prostrate stems extending over the surface of the sand. Its large, beautiful, pink and white flowers are sought after by wild-flower photographers. Other plants of this community include the prostrate, woody beach knotweed (*Polygonum paronychia*) which has very tiny flowers that bloom all year around; American dune-grass *(Elymus mollis)*, with its wide, dark green leaves and very large flower heads; American sea rocket *(Cakile edentula)*, with very small purple flowers, large seed pods and thick stems and leaves; and beach pea *(Lathyrus japonicus)*, a plant that looks similar to the common garden pea. The last two species are usually found quite close to the shore although occasionally will be found a distance inland.

Although these plants together make up the "pioneer community" of the moving sand habitat, it should be remembered that they may not all occur together at the same time. As few as one or two species of this group may be present on an area representing the "pioneer community" and beginning the successional process. Also, the species of this group may become established in and around the mounds caused by beach-grass and thus start the development of a stabilizing plant cover on these areas.

The end result of the establishment of these species on dry moving sand is to change the climate at the sand surface—there is protection from the winds, sand movement stops and temperatures are lower and moisture content higher because of shading by the plant leaves. This allows the establishment of plants that can tolerate little or no sand burial. Among the most common are the yellow-flowered seaside tansy *(Tanacetum camphoratum)*; the very abundant, purple-flowered seashore lupine *(Lupinus littoralis)* and a grass which is found in many different sand dune habitats, and many other places over the world, red fescue *(Festuca rubra)*. Not very important as far as succession is concerned but a very striking plant, is the sea-coast angelica *(Angelica hendersonii)*, a member of the carrot family that grows several feet tall on bluffs and dunes near the shore and has white flower heads up to six inches in diameter. On the parabola dune complex at Sand Lake, an extensive area of sand along the shore just south of the forest is being stabilized by a community that includes western bracken fern *(Pteridium aquilinum)* and pearly everlasting *(Anaphalis margaritacea)*.

After the low-growing, or herbaceous, vegetation has become well established, a community composed of various shrubs and tree seedlings begins to appear. These are plants which could not get started on open sand, but which easily become established in the shelter of the other plants. Quite frequently, by the time these shrubs and trees are seen, many of the plants which started the whole process are no longer apparent. Too much shade, no sand deposition, and competition from other plants, all play a part in this disappearance. These plants have literally forced themselves out of existence by the changes they brought about.

Two shrubs are most common on these areas: salal *(Gaultheria shallon)* and evergreen huckleberry *(Vaccinium ovatum)*. Along with them are usually found seedlings of the coast pine *(Pinus contorta)* and, in places well protected from the ocean winds, seedlings of Douglas-fir *(Pseudotsuga menziesii)*. This shrub stage develops rapidly into a forest such as is seen on the sand north and south of Florence, and on the dune ridges at Sand Lake. These forests are composed primarily of coast pine, and a varying, but small proportion of Douglas-fir. In low, moist areas of the older forests, it is possible to see saplings and larger trees of western hemlock (*Tsuga heterophylla*) and western red cedar (*Thuja plicata*).

The understory, or shrub, vegetation of these forests is usually quite thick and tangled, making it virtually impossible to walk through them except on trails. Dominant of these shrubs is western rhododendron (*Rhododendron macrophyllum*). Although generally 8 to 10 feet high, it sometimes grows into a small tree up to 25 feet in height. In late spring and early summer its clusters of large pink flowers brighten the shade of these forests.

Growing along with rhododendron is the evergreen huckleberry. It too grows tall in the shade of the forests—up to 8 feet. Its extremely small, bell-shaped flowers appear in early spring, and by early autumn the branches are loaded with the small, sweet berries. Salal is occasionally found as a tall spindly shrub in the forest but it attains its best growth and development at the edges of the forest in dense hedge-like stands in the open. The pinkish flowers are large and bellshaped. The fruit is sweet and edible, but not widely appreciated.

Also around the edges of the forest, or in cleared areas where there is neither full shade nor full sunlight, are found two other species which belong to the same family of shrubs as those already mentioned. One is the low growing, mat-forming bearberry or kinnikinnick (*Arctostaphylos uva-ursi*). Its bright red berries are an important source of food for wildlife. The other is an upright shrub, hairy manzanita (*Arctostaphylos columbiana*), characterized by hairy young twigs, and peeling bark on the older stems which reveals a smooth, red surface.

There is little or no plant life under the heavy shrub cover inside the forest but in places where, for one reason or another, the shrubs have been removed or have never become established, various forms of plant life can be seen. Chief among these are the lichens and mosses, simple plants with neither flowers nor true leaves, stems or roots. There are numerous species of these plants and they are difficult to identify. In deep forest shade is the most common moss (*Eurhynchium oreganum*), with finely branched stems and small, usually yellowish, "leaves." Most of the lichens are of the genus *Cladonia.* One species, which grows in light gray green tangled thread-like masses, is closely related to the "reindeer moss" of the Arctic tundra. Others are upright stalks, no more than two inches high, variously branched and colored. One of these, with a red tip resembling a matchstick, is commonly called "British Soldiers."

Another interesting plant which can be seen in these forests during the early summer is the small ground-cone *(Boschniakia hookeri),* a parasite on the roots of salal or other plants of the heath family. It sends a short (3 to 6 inches high) brownish-red stem above ground in spring. The leaves, which are scale-like, are without chlorophyll, and the flowers are pale and inconspicuous. The plant obtains all of its food for growth from the roots of the host plant (such as salal), without apparently, harming it.

Several more species of plants can be found in these sand dune forests, but those mentioned are the most numerous and obvious. Thus these forests are relatively simple in community composition as compared perhaps to the deciduous forests of the middle western or eastern states of the country. If they are not destroyed by new sand activity, the Douglas-fir will in time become dominant over the coast pine because of its longer life and larger size. Over long periods of time (hundreds of years) the Douglas-fir would be replaced by western hemlock and western red cedar because the seedlings of these species are able to develop readily in the shade of the Douglas-firs. Western hemlock and western red cedar thus become the climax species of the sand dune forests.

Another successional sequence takes place on sand dunes that is worthy of mention. In areas where there is little or no sand movement (the lee side of large masses of vegetation or in the troughs and sand banks left in the wake of a moving parabola dune) primary succession is started by red fescue growing as a bunch grass. Also found growing with the grass is sticky goldenrod *(Solidago simplex). The* scattered clumps of grass, each from two to five inches in diameter and spaced anywhere from 6 to 24 inches apart, along with clumps of goldenrod, effectively stop wind movement at the ground surface.

This altered ground environment permits the establishment of a moss, *Rhacomitrium canescens,* which tolerates warm, sunny, dry environments. Eventually the moss covers the ground like a thick carpet, even crowding out the red fescue and goldenrod. This, in turn, creates environmental conditions favorable for the establishment of seedlings of bearberry (kinnikinnick). The prostrate branches of this shrub spread over the moss carpet, eventually covering it and causing the disappearance of the shade intolerant moss.

As the moss dies off, seedlings of coast pine grow up through the bearberry. Again, if there is no disturbance through renewed sand activity, a coast pine sand dune forest would develop which would, in time, give way to the climax hemlock-cedar forest. One area in which this successional sequence is an important part of the dune landscape is the northeast end of the parabola dune complex at Sand Lake. It can be seen on both sides of Derrick Road where it crosses the dunes toward the Boy Scout camps.

Plant Communities
of the Deflation Plains

Unlike the moving sand dunes, the problem in the deflation plain environment is not one of sand deposition, but of an abundance of water. Many of the plants in these areas, then, are those which are adapted or able to grow in wet places, regardless of their location (on or off sand). However, the deflation plains are not uniformly wet, but rather there is a sort of gradation from the higher, dry edges to the very wet, lowest spots. Also there are hummocks and small areas of raised ground even in the lowest areas. The plants and plant communities show in a remarkable way these differences, sometimes very slight, in elevation of the ground surface in relation to the water table. Frequently as little as 6 inches vertical height will separate two rather distinct groupings of plants.

Because of these differences in environment due to the presence of water, it is possible to recognize four different plant communities on the deflation plains, each occurring in response to slightly different environmental conditions. Although there is some successional relationship among these, which will be discussed later, each can be considered an initial stage in primary succession on bare wet sand, leading ultimately to the establishment of a forest.

The Dry Meadow Community

Ecologically situated at the driest end of the four-community sequence, the dry meadow is almost exclusively dominated by three species: seashore lupine *(Lupinus littoralis)*, European beach-grass *(Ammophila arenaria)*, and seashore bluegrass *(Poa macrantha)*. Also found in the community, in varying amounts and in various places, are beach knotweed *(Polygonum paronychia)*, beach silver-top *(Glehnia littoralis)*, and gray beach pea *(Lathyrus littoralis)*. All these species have been encountered before in the discussion of the moving sand dune plant communities, making this particular group of plants a transitional community between sand and the deflation plain.

Of the species listed, seashore lupine is by far the most important, often occurring in dense stands. It is able to withstand only very slight amounts of sand deposition, and grows most vigorously in moist areas. European beach-grass, seashore bluegrass and the rest of the species listed, though well adapted to areas of active sand deposition, will become established and persist, in varying, but small amounts, in the more moist, more stable habitats.

The habitat of the community is dry compared with the rest of the deflation plain and becomes hummocky where the beach-grass clumps have collected sand. It usually shows quite a bit of bare ground and has the least number of different species of all the communities. It occurs on raised areas on the deflation plain, around the higher edges, and very typically just behind the foredune, where it gradually changes into a pure beach-grass community seaward as sand deposition increases. Sand deposition in the dry meadow varies from almost none where seashore lupine is found in dense stands, to considerable amounts in the more open stands of seashore bluegrass, beach silver-top and gray beach pea. The water table is more than three feet below the surface during the summer months, and water never stands on the surface during the wet winter months.

The community reaches its flowering peak in late spring and early summer. Particularly striking is seashore lupine which begins growth as early as December, is in full bloom by May, and generally is completely dried up by the end of August. Beach knotweed, on the other hand, has inconspicuous flowers during the entire year. Species present in quite small amounts (in addition to those already mentioned) include Australian fireweed *(Erechtites minima)*, sticky goldenrod *(Solidago simplex)* and salt rush *(Juncus lesueurii)*. The presence of salt rush indicates that the water table is, or at one time was, near the surface. This species tolerates considerable sand deposition, and will even form large sand mounds if there is a sufficient amount growing in one place.

The Meadow Community

A number of species are important in this community and are usually present, at least to some degree. They include two grasses, red fescue *(Festuca rubra)* and little hair-grass *(Aira praecox)*, false dandelion *(Hypochaeris radicata)*, coast strawberry *(Fragaria chiloensis)*, and seashore lupine. Also present, but in lesser amounts, are salt rush, pearly everlasting *(Anaphalis margaritacea)*, seaside tansy *(Tanacetum camphoratum)*, yarrow *(Achillea millefolium)*, various species of bentgrass *(Agrostis spp.)* and a number of species of moss. Both the total number of species and the thickness of the vegetation covering the ground increases considerably in this community.

Seashore lupine is abundant, but not quite as important here as in the dry meadow community. Its ability to grow under a relatively wide range of environmental conditions is shown by its presence in a number of different habitats. In some places, red fescue is the most important plant and may grow so thickly that a sod or turf is formed, but this is not always the case. The coast strawberry is usually quite common, but it grows low to the ground and is difficult to see, especially during the height of the growing season. In some places, it is found on moist sand banks or mounds or in almost pure stands and during mid-summer it is possible to pick fair quantities of the small, but sweet fruit. The little hair-grass is a very inconspicuous annual that produces a thick, lush growth during the spring months, but is completely dried up by early summer. The community is best seen during the summer months when it is characterized by the showy flowers of false dandelion, seaside tansy, pearly everlasting and yarrow.

The habitat is generally level, more moist than the previous community, and there is essentially no sand deposition. There is no specific zone of occurrence within the deflation plain, the distance to water table apparently being the chief factor. Where it has been measured, the water usually stands at between two and three feet below the surface during the summer months, and it will stand on the surface for one or two months during the wettest time of the year. There is frequently a layer of dry or decaying organic material on the surface of the sand under the dense cover of the lupine and grasses.

The Rush Meadow Community

This community is impressive for its dense growth of many species. Bare ground is hardly ever apparent. The heaviest growth in this community is by two species of rush, brown-headed rush *(Juncus phaeocephalus)* and sickle-leaved rush *(Juncus falcatus)*, and spring-bank clover *(Trifolium wormskjoldii)*. Also important and quite numerous are common California aster *(Aster chilensis)*, golden-eyed grass *(Sisyrinchium californicum)*, Pacific willow-herb *(Epilobium ciliatum)*, centaury *(Centaurium erythraea)* and several species of bent-grass *(Agrostis, spp.)*.

Where the clover and the rushes grow together, the clover is usually inconspicuous because it grows so close to the ground. In some places, however, the clover grows in almost pure stands, and in spring, the large pink flower heads make very colorful displays. The small, yellow flowers of the golden-eyed grass, a member of the iris family, and the pink flowers of willow-herb and centaury can be seen most of the summer. One of the last plants to flower is common California aster, and its purple flower heads can be seen late into the autumn.

A number of other species are found on these wet habitats, but some are important in only a few places while others are found everywhere, although only in small numbers. An example of the latter is, again, salt rush. This species is widely distributed in the deflation plains but hardly ever in great numbers in any one place. The twisted orchid *(Spiranthes romanzoffiana)* occurs quite commonly, but not in great numbers. Others which may grow in relatively large amounts in scattered locations include *Castilleja ambigua*, toad rush *(Juncus bufonius)* and green sedge *(Carex viridula)*. Two species in this category are especially interesting. The common monkey flower *(Mimulus guttatus)* grows in great numbers on the deflation plain that is located north and west of Lily Lake and just south of the outlet of Sutton Creek, north of Florence. It presents a striking appearance in late spring and early summer with its very red young leaves and large, showy yellow flowers. On the deflation plain straight west from Cleawox Lake one can find the sundew *(Drosera rotundifolia)*, the insectivorous plant that traps insects in a sticky excretion exuded from glandular hairs on the leaves. Although common in sphagnum bogs, it is not often found on the sand dune deflation plains.

The rush meadow community occurs on lower ground than the previous one. During the summer months, the water table is about 12 to 18 inches below the surface, and in the winter water will stand on the surface for three or four months. There is usually a thin layer of decomposing organic material on the surface of the sand, followed by up to an inch of a rusty-colored peat-like substance probably made up of the roots and rhizomes, both living and dead, of the many plants growing there.

The Marsh Community

These habitats are marshy in the sense that they are quite wet most of the year. In some places the water table is only 6 inches below the surface in the summer and in most places, kneeling or standing in one spot would produce moisture (except for the very driest periods). Water may actually stand on the surface for more than six months of the year.

The most characteristic species of this community is slough sedge *(Carex obnupta)* which is almost entirely restricted to these very wet places. Another species which is usually found here is Pacific silver weed *(Potentilla anserina).* Also preferring such wet places but present in lesser amounts are small creeping buttercup *(Ranunculus flammula),* king's gentian *(Gentiana sceptrum), Carex lenticularis,* and bog club-moss *(Lycopodium inundatum).* Salt rush *(Juncus lesueurii)* is also present.

During the summer, the yellow flowers of Pacific silver weed and creeping buttercup, and the blue flowers of the gentian provide the principal color of the community.

Each of these plant communities begins through the colonization of the different deflation plain habitats by the species that grow best in each particular environment. There is very little change from one community to the other because of the modification of growing conditions by the plants present. What often happens is that sand deposition or drainage reduces the amount of moisture on an area, and thus makes it possible for a drier habitat plant community to become established. On the other hand, drainage might be impeded (perhaps because of dune activity) so that conditions become wetter, and thus unfavorable for the plants present.

In any case, the actual course of succession is toward a shrub community and then to a forest, and this occurs very rapidly. The low, herbaceous communities provide the shelter needed for the development of the shrub and tree seedlings. The shrubs develop as two more or less distinct communities—dry and wet— although there is much intermixing. On the dry sites, one usually sees seedlings of evergreen huckleberry *(Vaccinium ovatum)* and salal *(Gaultheria shallon)*, while on wetter places, coast willow *(Salix hookeriana)* and wax myrtle *(Myrica californica)* are much more common. Seedlings of coast pine *(Pinus contorta)* are found in abundance in both, while those of Sitka spruce *(Picea sitchensis)* are fewer in number. The huckleberry and salal are both relatively low growing shrubs and the pines soon are much taller than they are. The willow and wax myrtle however, grow rapidly—as fast as the pines—and in a short time these tree-shrub communities are closed in with an almost impenetrable growth of vegetation.

An excellent example of a deflation plain in all stages of succession is found directly west of Cleawox Lake (Lane County). On this area, succession begins on the east, or inland edge, with the shrubs and trees getting progressively higher and the vegetation denser as one moves toward the ocean. It becomes so dense that one cannot move completely across the deflation plain from east to west. The east edge culminates in a fine stand of Sitka spruce. The seedlings of this tree cannot tolerate exposure to the severe climate of the open sand, so they are found only in the shelter of the taller shrubs and pines. The tops of the spruce trees usually begin to appear above the surrounding vegetation when they are about ten feet tall. From that point on, they grow faster than the pines, and if there are enough of them, will eventually dominate the forest community. The willows and wax myrtles will persist along with the pines for quite some time, growing into rather large, but much-branched trees. Eventually, however, the pines become the major plants of the forest. They in turn, as has been indicated, are eventually replaced by the larger and longer lived Sitka spruce, which is probably the climax forest species on these areas. A good example of a well-developed spruce forest can be seen near the junction formed by the highway that runs north from Cape Kiwanda (north of Pacific City), and the highway that goes north from Woods.

Because of the density of these forests, both pine and spruce stages, there is very little vegetation in the form of low shrubs or herbaceous plants under the forest canopy. Many types of mosses grow on the forest floor, and if there is any kind of an opening, or at the forest edge, many of the species of the deflation plain communities will be found. Extremely wet sites, such as the places where the Marsh Community is found, may have an open stand of coast pine under which is a thick growth of slough sedge.

The deflation plain west of Cleawox Lake is a good example of how rapidly a forest can develop in this type of environment. In 1915, a Forest Service worker described the area as "a grassy tract," apparently without any trees present.[27] Today some of the spruce are almost 50 feet tall, and nearly 12 inches in diameter near the base. Such rapid development can be traced on other deflation plains along the coast.

Sand Dune Stabilization Plantings

Sand dune stabilization (stopping sand movement) has been a subject of concern along the Oregon coast since the early 1900s. Sand blew into the rivers; it drifted across roads and highways, stopping traffic at certain times of the year; and occasionally even destroyed and buried houses. On the Clatsop area just south of the Columbia River, vast areas of sand become active when the natural vegetation which had stabilized the dune ridges was destroyed by human activities.

Early work on stabilization centered around the introduction of vast numbers of different kinds of plants from all over the world in an attempt to determine what would best hold the shifting sands in place. This was especially true of trees, and even today it is possible to find a few struggling survivors of those plantings. At the east edge of the deflation plain west of Cleawox Lake are the remnants of a 3.3 acre planting of maritime pine *(Pinus pinaster)* set out in 1915 and 1916. Several rows of eucalyptus trees were also planted, but none have survived. Unfortunately, the maritime pines are in the path of sand dunes invading the vegetated area, and soon may be completely lost.

The species finally decided upon for stabilization work were European beach-grass *(Ammophila arenaria)* and scotch broom *(Cytisus scoparius)*, both introduced from Europe; and coast pine *(Pinus contorta)* a native tree. Major planting efforts began in the 1930s but, except for the Clatsop area, plantings were incomplete and planning seems to have been somewhat haphazard. Dunes planted to beach grass alone do not become stabilized, but develop into large, unattractive grassy hummocks or mounds. Such an area is located just west of Sutton Lake, north of Florence, and can be reached from the Sutton Lake Campground.

The Clatsop area was successfully stabilized, however, and in the late 1940s attention was focused on the sand dunes between the Siuslaw and Siltcoos Rivers south of Florence. The first beach-grass plantings south of the Siuslaw were made in 1949 by the Bureau of Land Management and the Soil Conservation Service. At the Siltcoos River area, the U.S. Forest Service started its planting in 1957. Plantings have been made almost every year since that time in both locations. As of 1964, over 300 acres had been planted in the Siuslaw area; over 400 acres in the Siltcoos area. The general program followed in the stabilization work involves an initial planting of beach-grass, followed in a year or two by scotch broom, and then, either concurrently, or a year later, by coast pine.

The grass is planted as "sets" obtained from a beach-grass nursery. These sets are usually 2 years old when dug up, and are trimmed back to 22 inches in length. Planting is mostly by machine, the sets being spaced from 18 by 18 inches with five stems per hill to 12 by 12 inches with three stems per hill, the distance and number of stems depending mostly upon slope and exposure to wind. Dates of planting are generally from November 1 through April 1. The important factor is maximum temperature. Survival of beach-grass will be less than 50 percent if it is planted at temperatures of 55° or above. Fertilizers are applied to insure survival and vigorous growth of the newly planted grass. A nitrogen rate of 40 pounds per acre (200 pounds ammonium sulphate) seems to produce the best results. The fertilizer is applied at the start of spring growth, generally during March and April, after the winter rains have ended. This helps prevent loss of fertilizer which can be carried away from the soil surface by drainage water.

The secondary planting of scotch broom, a leguminous plant, is made as soon as the beach-grass forms a good cover, usually the year following the planting of the grass. One-year-old seedlings from nurseries are used, and they are spaced either 6 by 6 or 8 by 8 feet. Coast pine is planted either at the same time as the scotch broom or a year later. Two-year-old seedlings are used which are also planted at intervals of 6 by 6 or 8 by 8 feet. The availability of both funds and nursery stock for planting determines how much is planted each year.

The planting program is based on ecological principles of succession. The beach-grass is the primary planting and provides the initial stabilization of the sand surface. It also provides shelter for the scotch broom and coast pine seedlings. Since the beach-grass will deteriorate without a continuous supply of new sand, the fast growing scotch broom becomes important as a sand stabilizer and provides shelter for the smaller and slower growing pines. About 10 to 12 years after the initial planting, the pines begin to grow taller than the scotch broom. Since the scotch broom cannot tolerate shade, it will eventually disappear, to be replaced by native shrub species. The whole purpose of the planting is to produce a pine forest of the type present on undisturbed dunes, and results so far seem to indicate that this will take place.

Many of the yearly plantings can be seen from the road that turns off from U.S. Highway 101 between Florence and Honeyman State Park and goes out to the beach and the south jetty of the Siuslaw River. Some of the more recent plantings encountered just as the road moves over the dunes include a large bushy shrub that has masses of brilliant yellow flowers in late spring. This is the tree lupine *(Lupinus arboreus)*, a native of California. Plantings of various ages can also be seen around the U.S. Forest Service Recreation Area on the Siltcoos River.

References

1. Cooper, W.S. 1958. *Coastal Sand Dunes of Washington and Oregon.* Baltimore: Waverly Press, 169 p. (Geological Society of America Memoir 72)

2. *Ibid.*, p. 88

3. Tertiary events based on:

Snavely, Parke D., Jr., and Holly C. Wagner. 1963. Tertiary geologic history of western Oregon and Washington. Olympia: Department of Conservation, 25 p. (Report of Investigations 22)

For previous geologic history, sources consulted were:

Baldwin, Ewart M. 1950. *Geology of Oregon.* Ann Arbor, Michigan: Edwards Brothers, 136 p.

Cooper, W.S. 1958. *op. cit.*, p. 128-38.

Fenneman, Nevin M. 1931. *Physiography of the Western United States.* New York: McGraw-Hill, p. 458-65.

Smith, Warren D. 1933. Geology of the Oregon coast line. *Pan American Geologist* 59:33-44, 190-206, 241-58.

4. Hansen, Henry P. 1947. Postglacial forest succession, climate and chronology in the Pacific Northwest. *Transactions of the American Philosophical Society* 27:1-130.

5. Fairbridge, Rhodes W. 1960. The changing level of the sea. *Scientific American* 202(5):70-79.

6. Dicken, Samuel N. 1961. *Some Recent Physical Changes of the Oregon Coast.* Eugene: University of Oregon, Department of Geography, 151 p. (Report on Office of Naval Research Contract Nonr- 2771(04) Project NR 388-062)

7. Cooper, W.S. 1958. *op. cit.*, p. 13.

8. Decker, Fred W. 1961. *The Weather of Oregon.* 2nd ed. Corvallis: Oregon State University Press, 47 p. (Science Series Number 2)

9. Cooper, W.S. 1958. *op. cit.*, p. 17.

10. Twenhofel, William H. 1946. *Mineralogical and Physical Composition of the Sands of the Oregon Coast from Coos Bay to the Mouth of the Columbia River.* Portland: Department of Geology and Mineral Industries, 64 p. (Bulletin 30)

11. Berhardt, Paul. 1900. *Handbuch des Deutschen Dunenbaues.* Berlin: Paul Parey, 656 p.

12. Cooper, W.S. 1958. *op. cit.*, p. 27-49.

13. *Ibid.*, p. 49-64.

14. *Ibid.*, p. 131-38.

15. Hansen, Henry P. 1947. *op. cit.*

16. Cooper, W.S. 1958. *op. cit.*, p. 93.

17. Knowles, Margie Y. 1952. The Siuslaws. In: *The Siuslaw Pioneer.* Florence, Oregon: The Siuslaw Oar, p. 1-19.

18. Erlandson, L.D. 1947. The old dead trees. In: *The Siuslaw Pioneer.* Florence, Oregon: The Siuslaw Oar, p. 17.

19. Davidson, George. 1889. *Coast Pilot of California, Oregon and Washington.* Washington: U.S. Coast and Geodetic Survey. 718 p.

20. Dicken, Samuel N. 1961. *op. cit.*

21. Hanneson, Bill. 1962. Changes in the vegetation on coastal dunes in Oregon. Master's thesis. Eugene: University of Oregon. 103 numbered leaves.

22. Weaver, John E., and Frederic E. Clements. 1938. *Plant Ecology.* New York: McGraw-Hill. 601 p.

23. Shelford, Victor E. 1963. *The Ecology of North America.* Urbana: University of Illinois, 610 p.

24. House, H.D. 1914. The sand dunes of Coos Bay. *Plant World* 17:238-43. House, H.D. 1914. Vegetation of the Coos Bay region, Oregon. *Muhlenbergia* 9:81-100.

25. Byrd, N. L. 1950. Vegetation zones of coastal dunes near Waldport, Oregon. Master's thesis. Corvallis: Oregon State University. 44 numbered leaves.

26. Reardon, John Joseph. 1959. A study of the mammals of the dune environment of the Oregon coast with special reference to the adaptive behavior of Peromyscus maniculatus rubidus. Ph.D. thesis. Eugene: University of Oregon. 164 numbered leaves.

27. Mackechnie, F. L. 1915 and 1916. Report on sand dune planting experiments. Siuslaw National Forest. (Typed report on file in Supervisor's Office, Siuslaw National Forest, U.S. Forest Service, Corvallis, Oregon)

Identification of Sand Dune Plants

In order to provide a workable key the plants included here have been limited to those dune plants which are conspicuous or are of such general distribution as to attract attention. Mosses and lichens have been omitted, principally because of the difficulty involved in their identification. Other omissions will be found, and for those persons who are interested in the flora of the area and wish to get further information, the following publications will be useful:

Abrams, Leroy. 1940, 1950, 1951,1960. *Illustrated Flora of the Pacific States*. Stanford, California: Stanford University Press. 4 vols.
A flora of the vascular plants of Washington, Oregon, and California. Keys, descriptions, and illustrations.

Gilkey, Helen M., and LaRea J. Dennis. 1980. *Handbook of Northwestern Plants*. Corvallis, Oregon: OSU Bookstores Inc. 507 pages.
A flora of the vascular plants from the summit of the Cascades to the coastline of Washington and Oregon as far south as the Umpqua Divide. Keys, descriptions, and some illustrations.

Hickman, James C. (Ed.) 1993. *The Jepson Manual of Higher Plants of California*. Berkeley, California: University of California Press. 1,400 pages.
A flora of the vascular plants of California, with keys, descriptions, and some illustrations.

Hitchcock, C. Leo, Arthur Cronquist, Marion Ownbey and J.W. Thompson. 1955, 1959, 1961, 1964, 1969. *Vascular Plants of the Pacific Northwest*. University of Washington Publications in Biology, Vol. 17. Parts 1, 2, 3, 4 & 5.
A flora of the vascular plants of Washington, the northern half of Oregon, Idaho south of the Snake River Plains, the mountainous portion of Montana, and southern British Columbia. Keys, descriptions, and illustrations.

Hitchcock, C. Leo, and Arthur Cronquist. 1973. *Flora of the Pacific Northwest*. Seattle, Washington: University of Washington Press. 730 pages.
A condensed version of the above five volumes.

Lawton, Elva. 1971. *Moss Flora of the Pacific Northwest*. Nichinan, Japan: The Hattori Botanical Laboratory. 362 pages plus 195 plates.
Keys, descriptions, and illustrations.

McCune, Bruce, and Linda Geiser. 1997. *Macrolichens of the Pacific Northwest*. Corvallis, Oregon: Oregon State University Press. 386 pages.
Keys, descriptions, and photographs.

Peck, Morton Eaton. 1961. *A Manual of the Higher Plants of Oregon*. 2nd ed.
  Portland, Oregon: Binfords & Mort, Publishers. 963 pages.
  A descriptive flora with keys to the vascular plants of Oregon.

Using the Key

The purpose of a key is to enable the user to distinguish one thing from another on the basis of selected characteristics. It divides things (in this case, plants) into groups, breaks the large groups down to smaller groups, and then finally leads to the name of the plant in question. To do this the key gives a series of alternative choices, with only two choices at any particular point, eg. 1a or 1b. The opposing statements always are numbered alike but are lettered differently.

There are certain criteria to keep in mind in using a key. (1) You must read both choices carefully, for although the first choice sounds good, the second may be even better. (2) Be sure you understand the terms used and consult the glossary when necessary; guessing may result in making the wrong choice. (3) Once you have arrived at a name in the key, turn to the final section, read the description, and look at the photograph to be certain your unknown plant corresponds. In some cases additional species are listed below the principal one, and a further key is sometimes provided. If the plant does not match the given description, you have made a mistake in keying, or your plant may be one that was not included in this book.

Descriptive Section

There are ninety species in the key; these plus forty additional species are listed and described. Each of the ninety principal species is numbered to permit easy cross reference between the key and the descriptions. The arrangement of plants is by families, in a more or less natural sequence, and each family is briefly described. An attempt has been made to avoid technical botanical terminology; however, a certain number of terms is necessary, and definitions for these are in the glossary.

In addition to the names and descriptions of the plants, information is given on where they occur and what importance they have for humans. Their occurrence or distribution in the sand dune area is given first, followed by a general statement of their geographical range.

Scientific Names

The Latin or scientific name of a plant species consists of two words used together, the genus name followed by the specific name. This combination is followed by the name(s) or abbreviations of the name(s) of the botanist(s) responsible for establishing the specific name. Complete information of the rather complicated system of naming plants can be found in any of the following references:

Lawrence, George H.M. 1951. *Taxonomy of Flowering Plants.* New York: The Macmillan Co. 823 pages.

Radford, Albert E. 1986. *Fundamentals of Plant Systematics.* New York: Harper & Row. 498 pages.

Walters, Dirk R., and David Keil. 1995. *Vascular Plant Taxonomy.* 4th Ed. Dubuque, Iowa: Kendall/Hunt Publishing Co. 592 pages.

Common Names

Common names are given with the descriptions, but it must be remembered that, unlike scientific names, common names vary from place to place, and one common name may apply to two or more species or several different common names may be used for one species. Common names used here are considered to be the ones most often used in this area, and they may have little significance elsewhere.

Key to Sand Dune Plants

1a   Plants not producing seeds or flowers; reproducing by spores.

  2a   Leaves small and scale-like; sporangia borne in a terminal cone like structure .... (3) *Lycopodium inundatum*

  2b   Leaves large and variously divided; sporangia borne on the under surface of the leaves or on specialized leaf segments.

    3a   Leaves of two types, fertile (spore-bearing) leaves erect, sterile ones spreading, linear or lance-shaped in outline, once divided .... (1) *Blechnum spicant*

    3b   Leaves all alike, much divided, triangular in outline ....
    .... (2) *Pteridium aquilinum*

1b   Plants producing seeds and flowers or seed-bearing cones.

  4a   Trees or shrubs.

    5a   Trees with linear, needle-like or scale-like leaves.

      6a   Leaves scale-like; bark reddish-brown, shredding ....
      .... (7) *Chamaecyparis lawsoniana*

      6b   Leaves linear or needle-like; bark not shredding.

        7a   Leaves borne in 2's, surrounded at the base by a sheath
        .... (5) *Pinus contorta*

        7b   Leaves borne singly.

          8a   Leaves 4-angled, stiff and sharp-pointed ....
          .... (4) *Picea sitchensis*

          8b   Leaves flattened, not sharp-pointed ....
          .... (6) *Pseudotsuga menziesii*

    5b   Shrubs (if tree-like, then leaves never linear, needle-like or scale-like).

      9a   Petals absent; flowers in catkins.

        10a   Leaves opposite .... (55) *Garrya elliptica*

        10b   Leaves alternate.

          11a   Leaves gland-dotted; fruit waxy-coated, berry-like
          .... (27) *Myrica californica*

          11b   Leaves not gland-dotted; fruit a capsule ....
          .... (26) *Salix hookeriana*

9b Petals present; flowers not in catkins.

    12a   Stems at least partly prickly or spiny.

        13a   Flowers yellow; leaves much reduced; spines stiff .... (46) *Ulex europaea*

        13b   Flowers red; leaves large, divided into 3 leaflets .... (39) *Rubus spectabilis*

    12b   Stems without prickles or spines.

        14a   Flowers large and showy, $1^{1}/_{4}$-$2^{1}/_{2}$ inches wide; fruit a dry capsule .... (60) *Rhododendron macrophyllum*

        14b   Flowers much smaller, less than 1 inch wide.

          15a   Flowers yellow.

            16a   Flowers pea-shaped; leaves compound.

                17a   Leaflets 3; flowers in clusters of 2-3 .... (40) *Cytisus scoparius*

                17b   Leaflets 7-11; flowers in dense racemes.............. .... (43) *Lupinus arboreus*

            16b   Flowers not pea-shaped; leaves simple.

                18a   Leaves opposite; fruit a berry .... (75) *Lonicera involucrata*

                18b   Leaves alternate; fruit a downy achene................ .... (81) *Baccharis pilularis*

          15b   Flowers blue or white to pink.

            19a   Flowers blue or bluish-tinged.

                20a   Flowers bright blue; leaves simple........................ .... (48) *Ceanothus thyrsiflorus*

                20b   Flowers bluish-tinged; leaves palmately compound .... (43) *Lupinus arboreus*

            19b   Flowers white to pink.

                21a   Petals free from each other or nearly so; margins of the leaves rolled under .. (59) *Ledum glandulosum*

                21b   Petals united into an urn-shaped corolla.

22a Leaves mostly over 2 inches long, minutely toothed; branches zig-zag near the tip .... (58) *Gaultheria shallon*

22b Leaves less than 2 inches long; branches not zig-zag.

  23a Leaves toothed; fruit black .... (61) *Vaccinium ovatum*

  23b Leaves entire; fruit red or orange.

    24a Young branches with numerous bristly hairs; leaves gray-green .... (56) *Arctostaphylos columbiana*

    24b Young branches not bristly; leaves green .... (57) *Arctostaphylos uva-ursi*

4b Herbs.

  25a Plants with parallel-veined leaves that are simple and entire; flower parts usually in 3's (rarely in 2's or 4's, but never 5's), or perianth parts reduced or absent.

    26a Perianth present; fruit several-seeded.

      27a Perianth inconspicuous, dry, brown or greenish.

        28a Plants apparently leafless, with slender, green, unbranched stems .... (21) *Juncus lesueurii*

        28b Plants with normal leaves.

          29a Leaves in a basal cluster; flowers in a spike terminating a naked stem.................................... .... (74) *Plantago maritima*

          29b Leaves arranged along the stem; flowers in spherical terminal clusters...... (20) *Juncus falcatus*

      27b Perianth conspicuous, white or brightly colored.

        30a Flowers regular (perianth parts of equal size and shape).

          31a Leaves heart-shaped; flowers white; perianth parts 4 .... (22) *Maianthemum dilatatum*

          31b Leaves narrow and grass-like; perianth parts 6..... .... (23) *Sisyrinchium califomicum*

        30b Flowers irregular.

          32a One perianth part forming a spur; spike not twisted .... (24) *Piperia elegans*

          32b No perianth parts forming spurs; spike twisted .... (25) *Spiranthes romanzoffiana*

26b Perianth parts absent or represented by bracts, bristles or linear scales; fruit 1-seeded.

  33a Stems solid or filled with a spongy pith; leaves 3-ranked.

    34a Leaves reduced to basal sheaths.

      35a Plants 2 feet tall or less; spike terminal .... (17) *Eleocharis macrostachya*

      35b Plants 3-15 feet tall; inflorescence of several spikelets .. .. (19) *Scirpus validus*

    34b Stems leafy.

      36a Spike solitary with numerous long silky hairs (appearing as a cottony ball) .... (18) *Eriophorum chamissonis*

      36b Spikes without long silky hairs.

        37a Inflorescence a large terminal head .... (15) *Carex macrocephala*

        37b Inflorescence of several small spikes .... (16) *Carex obnupta*

  33b Stems hollow except at the nodes; leaves 2-ranked.

    38a Sweet-scented grasses; spikelets with 2 empty lemmas below the floret .... (11) *Anthoxanthum odoratum*

    38b Not sweet-scented; spikelets without empty lemmas below the floret.

      39a Spikelets 1-flowered.

        40a Large stout grasses (2-6 feet tall) with long creeping rootstocks; inflorescence dense, cylindrical 4-12 inches long .... (10) *Ammophila arenaria*

        40b Small grasses; inflorescence if cylindrical, less than 4 inches long .... (8) *Agrostis stolonifera*

      39b Spikelets 2- to many-flowered.

        41a Spikelets sessile; spikes cylindrical; plants robust, 2-6 feet tall .... (12) *Elymus mollis*

        41b Spikelets not sessile; plants usually much smaller.

42a Spikelets 2-flowered; glumes longer than the lemmas ....
.... (9) *Aira praecox*

42b Spikelets 3- to many-flowered.

    43a Lemmas keeled, awnless .... (14) *Poa macrantha*

    43b Lemma rounded on the back, awned ....
.... (13) *Festuca arundinacea*

25b Plants with netted-veined leaves, or in linear leaves a single longitudinal vein; flower parts usually in 5's (rarely in 3's or 4's).

    44a Leaves compound.

        45a Flowers small, borne in involucrate heads or dense umbels.

            46a Leaves divided into more or less definite leaflets; flowers in dense umbels .... (54) *Sanicula arctopoides*

            46b Leaves divided into many small segments; flowers in heads.

                47a Plants densely silky or woolly throughout; leaves divided into linear segments ....
.... (79) *Artemisia pycnocephala*

                47b Plants not densely silky or woolly; leaves finely divided, the segments not linear.

                    48a Heads yellow .... (90) *Tanacetum camphoratum*

                    48b Heads white or pinkish ... (76) *Achillea millefolium*

        45b Flowers various, but not borne in involucrate heads or dense umbels.

            49a Corolla strongly irregular.

                50a Leaves palmately compound.

                    51a Leaflets 3 .... (45) *Trifolium wormskjoldii*

                    51b Leaflets 5-8 .... (44) *Lupinus littoralis*

                50b Leaves pinnately compound.

                    52a Flowers 1-2 inches long, yellow, the throat usually red-spotted .... (72) *Mimulus guttatus*

                    52b Flowers less than 1 inch long,

                        53a Leaflets 16-30 .... (47) *Vicia gigantea*

                        53b Leaflets 2-10.

54a   Plants silvery; flowers $1/2$-$3/4$ inches long ....
.... (42) *Lathyrus littoralis*

54b   Plants green; flowers $3/4$ inch long or more ....
.... (41) *Lathyrus japonicus*

49b   Corolla regular.

55a   Stems creeping, rooting at the nodes; stamens numerous.

56a   Fruit fleshy (a strawberry); flowers white; leaflets 3 ......
.... (37) *Fragaria chiloensis*

56b   Fruit dry (achenes); flowers never white; leaflets 3-31 ..
.... (38) *Potentilla anserina*

55b   Stems not creeping and rooting at the nodes; stamens fewer than 10.

57a   Flowers blue.

58a   Leaves finely divided; flowers approximately $1/4$ inch long .... (67) *Gilia capitata*

58b   Leaves divided into well-defined leaflets; flowers $1/2$-$3/4$ inch long .... (69) *Phacelia bolanderi*

57b   Flowers white.

59a   Leaves silvery-pubescent on both surfaces; petioles not sheathing; fruit not winged ... (68) *Phacelia argentea*

59b   Leaves green above, pubescent beneath; leaves with sheathing petioles; fruit broadly winged.

60a   Plants prostrate or creeping, less than 6 inches tall .... (53) *Glehnia littoralis*

60b   Plants erect, over 1 foot in height ....
.... (52) *Angelica hendersonii*

44b   Leaves all simple, entire to variously toothed, lobed or cleft.

61a   Leaves opposite.

62a   Flowers irregular, bright yellow usually with red spots in the throat.. (71) *Mimulus guttatus*

62b   Flowers regular.

63a Herbage succulent.

  64a Flowers axillary, white or pink .... (63) *Glaux maritima*

  64b Flowers in umbels or in heads.

    65a Leaves narrow, not glandular; flowers in involucrate heads .... (87) *Jaumea carnosa*

    65b Leaves ovate to nearly orbicular, glandular on the under surface; flowers in umbels .... (30) *Abronia latifolia*

63b Herbage not succulent.

  66a Flowers rose-colored to purplish.

    67a Flowers 1 inch long or more, purplish blue .... (65) *Gentiana sceptrum*

    67b Flowers less than $^3/_4$ inch long; rose colored or purplish-red.

      68a Leaves entire, with 3 principal veins from the base ... . (64) *Centaurium erythraea*

      68b Leaves sharply and minutely toothed; pinnately veined .... (51) *Epilobium ciliatum*

  66b Flowers white or yellowish-orange.

    69a Leaves small, spine-tipped; stems prostrate; plants forming dense mats on sand .... (31) *Cardionema ramosissima*

    69b Leaves not spine-tipped; stems erect or decumbent.

      70a Flowers white; leaves acute .. (32) *Cerastium arvense*

      70b Flowers yellow-orange; leaves obtuse .... (49) *Hypericum anagalloides*

61b Leaves alternate or all basal.

  71a Leaves all basal.

    72a Leaves tubular, pitcher-shaped; insectivorous plants .... (35) *Darlingtonia californica*

    72b Leaves not tubular or pitcher-shaped.

73a Leaves with numerous gland-tipped hairs; herbage reddish; insectivorous plants .... (36) *Drosera rotundifolia*

73b Leaves not glandular; herbage green.

74a Leaves linear, grass-like, entire .... (62) *Armeria maritima*

74b Leaves not linear; margins deeply toothed or lobed.

75a Flowering stalk unbranched; heads solitary .... (88) *Leontodon taraxacoides*

75b Flowering stalks branched; several-headed .... (86) *Hypochaeris radicata*

71b Leaves not all basal.

76a Leaves deeply cleft or deeply lobed.

77a Flowers white or purplish .... (34) *Cakile edentula*

77b Flowers yellow.

78a Leaves, at least the lower, long petioled ....(54) *Sanicula arctopoides*

78b Leaves sessile and sheathing .. (82) *Cotula coronopifolia*

76b Leaves entire or toothed.

79a Flowers in involucrate heads; individual flowers small.

80a Ray flowers present, yellow or white to purplish-blue.

81a Ray flowers yellow; heads very small .... (89) *Solidago simplex*

81b Ray flowers purplish-blue to white.

82a Involucral bracts in 1 or 2 series; plants less than 1 foot tall ....(84) *Erigeron glaucus*

82b Involucral bracts overlapping in several series; plants 1-3 feet tall .... (80) *Aster chilensis*

80b Ray flowers absent.

83a Leaves all entire.

84a Involucral bracts white and papery .... (78) *Anaphalis margaritacea*

84b Involucral bracts purplish or brownish .... (85) *Gnaphalium purpureum*

83b Leaves, at least the lower, toothed or shallowly lobed.

85a Leaves sharply and finely toothed; annual ....
.... (83) *Erechtites minima*

85b Leaves coarsely toothed, perennials.

86a Leaves silvery-pubescent ..... (77) *Ambrosia chamissonis*

86b Leaves glabrous .... (81) *Cotula coronopifolia*

79b Flowers not in involucrate heads.

87a Flowers irregular.

88a Plants cone-shaped, entirely reddish-brown; leaves reduced to scales; root-parasites .... (73) *Boschniakia hookeri*

88b Plants not cone-shaped, green; leaves not reduced to scales.

89a Plant 1 foot tall or less; flowers light yellow with purple spots.. (70) *Castilleja ambigua*

89b Plants 2 to 9 feet tall; flowers white or rose-purple ......
.... (71) *Digitalis purpurea*

87b Flowers regular.

90a Leaves with sheathing stipules.

91a Flowers pink; leaves broadly linear or oblong, the margins entire .... (28) *Polygonum paronychia*

91b Flowers green; leaves oblong to lance-shaped, the margins crisped.......................... (29) *Rumex maritimus*

90b Leaves not sheathing at the base.

92a Flowers approximately 2 inches across, trumpet-shaped, pink to rose-purple; stems trailing ....
.... (66) *Calystegia soldanella*

92b Flowers smaller, not trumpet-shaped.

93a Herbage pubescent, grayish or silvery.

94a Stems erect; leaves silvery, flowers white ....
.... (68) *Phacelia argentea*

94b Stems prostrate, leaves gray; flowers yellow, turning red in age .... (50) *Camissonia cheiranthifolia*

93b Herbage glabrous, green.

    95a Plants with stolons; flowers yellow; herbage not succulent ...
    .... (33) *Ranunculus flammula*

    95b Plants lacking stolons; flowers white to pink or purple; herbage more or less succulent.

        96a Leaves entire; flowers axillary .... (63) *Glaux maritima*

        96b Leaves coarsely toothed; flowers in a raceme ....
        .... (34) *Cakile edentula*

Description of Sand Dune Plants

Ferns

Herbaceous plants lacking flowers and reproducing by spores rather than seeds; spores produced in specialized structures called sporangia. In the Polypodiaceae (Fern family), clusters of sporangia can frequently be seen as tiny raised, brown dots or lines on the undersides of the leaves.

1. *Blechnum spicant (L.)* Roth.

(=*Struthiopteris spicant* (L.) Weis.)
Deer-fern
Polypodiaceae (Fern family)
Leaves of two types, the sterile ones spreading or ascending, the fertile ones erect; leaves all once divided, linear or lance-shaped in outline, the segments of the fertile leaves narrower than those of the sterile ones.

Damp forests, moist meadows and moist areas in the deflation plains. Alaska to California and in Eurasia.

2. *Pteridium aquilinum* (L.) *Kuhn var. pubescens Underw.*
Western bracken fern
Polypodiaceae (Fern family)
Large ferns with creeping rhizomes; leaves triangular in outline, one to six feet long, three times divided, segments with inrolled margins under which are borne the sporangia.

Common on the dunes and in fields and woods. Throughout western North America. This species often poisons horses and cattle, causing a disease known as "fern-staggers." The young shoots, while still curled, have been boiled and eaten as a substitute for asparagus.

In addition to these two common ferns several other members of the Polypodiaceae occasionally occur on the dunes. *Polypodium vulgare L.* (licorice fern) is sometimes found in the sand, but more often is seen growing on rocks, logs and mossy tree trunks. The rhizomes of this species have a licorice taste; the leaves are once divided nearly to the midrib, the segments are alternate with broad confluent bases, the margins of the segments are finely toothed. *Polystichum munitum* (Kaulf.) Presl (sword fern), though more common inland, is occasionally found in abundance in the moist areas of the deflation plains. The leaves are lance-shaped in outline, once divided into narrow segments, each segment with a small lobe at the base.

A rare but unique inhabitant of the wet areas of the dunes and of Sphagnum bogs is *Botrychium multifidum* (Gmel.) Rupr. (leathery grape-fern). This fern is a member of the Ophioglossaceae (Adder's-tongue family), a family characterized by the existence of a sterile leaf segment and a spore-bearing segment rather than having sporangia borne on the backs of the leaves as in Polypodiaceae. The sterile segment of the leaf of the leathery grape-fern is thick and almost fleshy in texture, is much divided and is borne below the fertile portion which is long-stalked and diffusely branched.

Club-mosses

Members of this group belong to the family Lycopodiaceae. They somewhat resemble mosses in appearance and have numerous small, linear, scale-like leaves. They reproduce by spores produced in sporangia borne in the axils of the leaves, or in the axils of bracts in cone-like structures.

3. *Lycopodium inundatum* L.
Bog club-moss
Lycopodiaceae (Club-moss family)
Stems creeping; leaves linear, overlapping; stems terminated by a cone-like structure $1/2$ to 2 inches long.
    Coastal bogs. Alaska to Oregon.

Conifers

Mostly evergreen trees with linear or scale-like leaves. Conifers reproduce by seeds, which are usually borne naked on the scales of woody cones.

4. *Picea sitchensis* (Bong.) Carr.
Sitka spruce
Pinaceae (Pine family)
Evergreen tree up to 180 feet tall; leaves 4-angled, $^1/_2$ to 1 inch long, stiff, spirally arranged, their woody peg-like bases remaining on older twigs; cones 2-4 inches long.

Deflation plains and wet places on dunes, frequently with coast pine. Alaska to California.

5. *Pinus contorta* Loud.
Coast pine, beach pine or lodgepole pine
Pinaceae (Pine family)
Small tree up to 60 feet tall, but usually much lower; bark dark-colored; leaves needle-like, 1-3 inches long, 2 leaves to a bundle; cones $1-1^3/_4$ inch long.

Forming dense forests on the deflation plains; a pioneer species on almost any site except moving sand. Alaska to California.

Indians reportedly used the inner bark, fresh or pulverized and dried, as emergency food.

6. *Pseudotsuga menziesii* (Mirb.) Franco
Douglas-fir
Pinaceae (Pine family)
Large trees up to 200 feet tall; leaves linear, spirally arranged; cones up to $4^1/2$ inches long with the bracts projecting from between the scales.

On drier dune ridges, sometimes with coast pine or western hemlock. British Columbia to California and the Rocky Mountains. This is the most important lumber tree in the Northwest.

In addition to the above species, *Tsuga heterophylla* (Raf.) Sarg. (western hemlock), also of the Pinaceae, can occasionally be found on stable dune ridges with Douglas-fir and coast pine. Western hemlock is a graceful appearing tree with slender drooping branchlets; the leaves are $1/4$ to $3/4$ of and inch long, spirally arranged, but appearing 2-ranked by twisting; cones $3/4$ to $1^1/2$ inch long.

7. *Chamaecyparis lawsoniana* Parl.
Port Orford cedar
Cupressaceae (Cypress family)
Trees up to 90 feet in height; bark reddish-brown, deeply grooved and shredding; leaves scale-like; fruiting cone round, about $3/8$ of an inch in diameter, made up of 7-10 shield-shaped scales.

Mainly in the southern coastal areas. Oregon to northern California. Widely planted as an ornamental.

Another member of the Cupressaceae, *Thuja plicata* D. Don (western red cedar or giant cedar), can be found in wet interdune depressions. It is a large tree up to 225 feet tall with scale-like leaves (about $1/8$ of an inch long); cones $1/4$ to $1/2$ inch long, made up of 4-6 pairs of woody scales.

Flowering Plants

Herbs, shrubs or trees; flowers typically composed of sepals, petals, stamens and pistil (see diagrams). Any one or more of these parts may be missing, and the sepals and petals are occasionally variously modified.

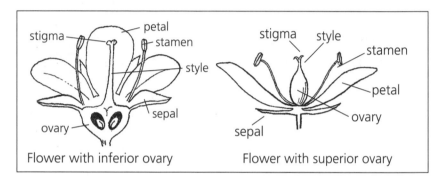

Flower with inferior ovary   Flower with superior ovary

Gramineae (Grass family)

The Grass family, from the standpoint of numbers of individuals, is the largest and most widely distributed of the families of vascular plants, and because of the many species used for food for man and for his domestic animals, it is probably of greater economic importance than any other family. Notable among the food crops are corn, barley, rice, rye and wheat. Because the family is very large and morphologically complex, it has acquired a terminology peculiar to itself. The basic unit of the inflorescence is the *spikelet*. A typical spikelet consists of a short axis on which the flowers (florets) are borne, subtended at the base by a pair of empty bracts (*glumes*). Each spikelet is composed of one to many florets and the glumes. Each floret is composed of 2 bracts, enveloping the 3 (rarely 1-6) stamens and the pistil; the first and usually the largest bract is the lemma, and the second is the *palea*. The vegetative structures of the grasses are also distinctive; the jointed stem is made up of nodes and internodes, the nodes are typically solid and the internodes hollow. The leaves are borne on the stem in two ranks, one leaf at each node; they are parallel-veined and are composed of two parts, the *sheath* which surrounds the stem, and the typically flat portion, the *blade,* The region on the back of the leaf at the union of the sheath and the blade is called the *collar,* and the area of union on the inner margin usually has a membranaceous or fringed appendage called the *ligule.*

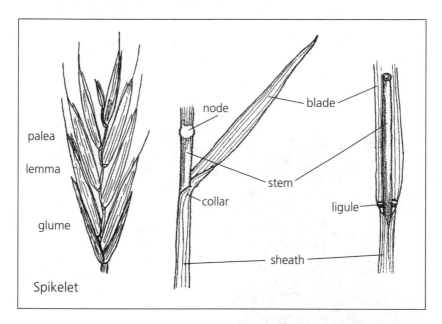

8. *Agrostis stolonifera* L.
Creeping bent-grass
Stems as much as 3 feet in length, creeping and rooting at the nodes; inflorescence delicate, the branches short; spikelets one-flowered.

Common in moist areas of the dunes. Washington to California.

Three other species of *Agrostis*, *A exarata* Trin. (western bent-grass), *A. oregonensis* Vas. (Oregon bent-grass) and *A. pallens* Trin. (seashore bent-grass), commonly occur on the deflation plains in areas of little or no sand deposition, or sometimes as pioneers at the edges of the deflation plains.

The four species of *Agrostis* may be separated as follows:

la. Plants without creeping rhizomes.

    2a. Inflorescence widely branched .... *A. oregonensis*

    2b. Inflorescence narrow and dense .... *A. exarata*

lb. Plants with creeping rhizomes.

    3a. Palea minute or wanting .... *A. pallens*

    3b. Palea at least half as long as the lemma .... *A. stolonifera*

9. *Aira praecox* L.
Little hair-grass

Slender annual; stem 2 to 10 inches tall; leaf blades short; inflorescence dense and narrow; spikelets 2-flowered.

Dry deflation plains with seashore lupine. British Columbia to California and on the Atlantic Coast; introduced from Europe.

Similar to *A. praecox*, but with an open inflorescence is *Aira caryophyllea* L. (silvery hair-grass). The species occur in similar habitats and sometimes together.

10. *Ammophila arenaria* (L.) Link

Beach-grass or European beach-grass

Stout perennial with long scaly rhizomes; flowering stems 2-6 feet tall; blades involute; inflorescence cylindrical, dense; spikelets 1-flowered, $3/8$ to $5/8$ inch long.

Forms the foredune; used as a sand binder in dune stabilization work.

California to Washington; introduced from Europe.

Ammophila breviligulata Fern. (American beach-grass) has been reported on the dunes in northern Oregon. It differs in having ligules only $1/8$ of an inch long or less, whereas they are $3/8$ to over 1 inch long in *A. arenaria.*

11. *Anthoxanthum odoratum* L.

Sweet vernal grass

Sweet-scented perennial, 8-24 inches in height; inflorescence nearly cylindrical; spikelets with one small terminal fertile floret and two larger sterile lemmas; whole plant pale yellowish-green in color.

Frequent on the dunes and along roadsides, from Europe.

12. *Elymus mollis* Trin.

Sea lyme-grass or American dune-grass

Robust perennial, 2-6 feet tall; leaf blades blue-green, 1/4 to 1/2 inch wide, not involute; inflorescence cylindrical, dense and soft; spikelets sessile, usually paired.

Mostly on the foredunes associated with beach grass. Alaska to California, the Great Lakes and on the Atlantic Coast.

This species of E*lymus* is sometimes put in the genus *Leymus*.

13. *Festuca arundinacea* Schreber

Tall fescue or alta fescue

Stout perennial, $1^1/2$ to 4 feet tall; leaf blades flat, $^1/8$ to $^1/2$ inch wide; inflorescence narrow with short appressed branches; spikelets several-flowered.

Established in the northern and central coastal counties and inland; introduced from Europe.

One other species of *Festuca* is common in the coastal region. *F. rubra* L. (red fescue), a perennial 1 to $3^1/2$ feet tall with reddish decumbent stems and leaf blades folded or involute, less than $^1/8$ inch wide, is common on the dry deflation plains where there is little or no sand deposition; it grows as a bunch grass on the drier areas such as at Sand Lake but is more common inland.

14. *Poa macrantha* Vas.
Seashore bluegrass
Perennial with long
rhizomes; stems decumbent,
8-16 inches tall; leaf blades
stiff, involute; inflorescence
dense; spikelets $3/8$ to $3/4$
inch long.

Beaches and dunes.
Washington to California.

Poa confinis Vas. (dune
bluegrass) also inhabits the
dunes and the sandy coastal
meadows. it is usually less
than 6 inches tall and the spikelets are about $1/8$ of an inch
long.

Cyperaceae (Sedge family)
The Sedge family is large and quite complex but is of little
economic importance; it is, however, important as a source of
food and shelter for wild fowl and various game animals. The
stems are typically solid and often triangular in cross-section.
Leaves, when not all basal, are 3-ranked. The leaf consists of a
blade and a closed sheath. The flowers are small, are subtended
by a chaffy bract or by bristles, and are arranged in spikelets;
the fruit is an achene.

15. *Carex macrocephala* Willd.
Large-headed sedge
Perennial; stems 1-12 inches tall,
arising from long horizontal
rhizomes; leaves $1/8$ to $1/4$ inch
wide, exceeding the stem in
length; head terminal, large and
dense.

Areas of sand movement and
deposition. Alaska to Oregon;
China and Japan.

16. *Carex obnupta* Bail.

Slough sedge

Perennial; stems somewhat tufted from stout creeping rhizomes; leaf blades $1/8$ to $1/4$ inch wide, stiff, channeled, keeled near the base; inflorescence of several cylindrical dark brown spikes, these often drooping.

Wet deflation plains with willows or under coast pine. British Columbia to California.

Two other species of *Carex* occur in abundance in the coastal area: *C. livida* (Wahl.) Willd. (livid sedge) is found in sphagnum bogs from Alaska to California, and *C. pansa* Bail. (sand dune sedge) is found on the high beaches and dunes from Washington to California. Two additional species are not at all common but may be locally abundant in wet places on the deflation plains; these are *C. lenticularis* Michx. var. *limnophila* (Holm.) Cronq. and *C. viridula* Michx. (green sedge).

The species of *Carex* may be separated as follows:

1a. Stigmas 2.

    2a. Spikes crowded into a short head .... *C. pansa*

    2b. Spikes separate.

        3a. Spikes sessile .... *C. lenticularis* var. *limnophila*

        3b. Spikes stalked .... *C. obnupta*

1b. Stigmas 3.

    4a. Spikes crowded into a large terminal head .... *C. macrocephala*

    4b. Spikes long and cylindric.

        5a. Stems shorter than the leaves .... *C. viridula*

        5b. Stems equalling or longer than the leaves .... *C. livida*

17. *Eleocharis macrostachya* Britt.
Creeping spike-rush
Stems 4 to 24 inches tall from stout creeping rhizomes, round in cross-section; leaves bladeless, the sheaths all basal; spike terminal, lance-shaped. Marshy areas in the dunes and inland.

Throughout most of the northern United States and Southern Canada.

18. *Eriophorum chamissonis* C.A. Mey
Russet cotton-grass
Perennial 1-$3^1/_2$ feet tall, with slender rhizomes; stems triangular in cross-section, leafy; spike terminal, solitary, bearing numerous long yellow-brown silky hairs.

Coastal bogs. Alaska to Oregon and on the Atlantic coast.

19. *Scirpus validus* Vahl.
American great bulrush
Robust perennial with short stout rhizomes; stems 3 to 15 feet tall; leaves bladeless or nearly so, the sheaths all basal; stems round in cross-section; inflorescence of numerous spikelets, these usually in small clusters.

Swamps and wet meadows. British Columbia to California and on the Atlantic coast.

Juncaceae (Rush family)

A relatively small family of little economic importance, although a few species of rushes are used as ornamentals in aquatic habitats. Stems commonly leafy, the leaves consisting of a blade and a sheath or sometimes reduced to sheaths only; flowers small, consisting of 6 chaffy segments; stamens 3 or 6; fruit a capsule with few to many seeds.

20. *Juncus falcatus* E. Mey.
Sickle-leaved rush
Perennial; rhizomes long and scaly; leaves mostly basal, thick; heads solitary or rarely 2 or 3, 5- to 25-flowered.

Wet deflation plains, associated with brown-headed rush. British Columbia to California; Japan and Australia.

21. *Juncus lesueurii* Boland.
Salt rush
Perennial; stems stiff and wiry, 1-3$^1/_2$ feet tall, rising from long creeping rhizomes; leaves reduced to basal sheaths; inflorescence usually a dense laterally appearing head.

Very common in all wet places on dunes, sometimes forming thick stands. Washington to California.

The following species of *Juncus* are also common in wet areas in the coastal region: *J. acuminatus* Michx. (sharp-fruited rush); *J. articulatus* L. (jointed-leaved rush); *J. bolanderi* Engelm. (Bolander's rush); *J. bufonius* L. (toad rush) *J. supiniformis* Engelm. (hair-leaved rush), and *J. phaeocephalus* Engelm. (brown-headed rush).

The species of *Juncus* can be separated as follows:

la.   Leaves reduced to sheaths; inflorescence appearing to be lateral .
.... *J. lesueurii.*

lb.   Leaves with well developed blades; inflorescence not lateral.

    2a.   Flowers borne singly or in clusters of 2 or 3; annual ....
.... *J. bufonius*

    2b.   Flowers borne in 5- to many-flowered heads; perennials.

        3a.   Leaves without partitions (i.e. internal septa) .. *J. falcatus*

        3b.   Leaves with numerous partitions (internal septa).

            4a.   Leaves strongly flattened with one edge turned toward the stem .... *J. phaeocephalus*

            4b. Leaves not strongly flattened.

                5a.   Stamens 3.

                    6a.   Heads 1-8, densely clustered; dark brown ....
.... *J. bolanderi*

                    6b.   Heads many (10-50), loosely clustered; light brown .... *J. acuminatus*

                5b.   Stamens 6.

                    7a.   Stems weak, decumbent; perianth segments about $3/16$ of an inch long .... *J. supiniformis*

                    7b.   Stems stout, erect; perianth segments less than $1/8$ inch long .... *J. articulatus*

Liliaceae (Lily family)

This is a large and widely distributed family, but with few representatives in the coastal area. The family is important economically for its large number of ornamentals, including hyacinths, lilies and tulips, and for such vegetable crops as onions and asparagus. The distinguishing characteristics of the family are the usually showy flowers with flower parts in 3's or rarely 4's, the leaves usually parallel-veined, and the superior ovary.

22. *Maianthemum dilatatum* (Wood) Nels. & Macbr.
False lily-of-the-valley
Stems up to 12 inches high from horizontal rhizomes; leaves 2 or 3, nearly heart-shaped; flowers small, white, several to many in a narrow inflorescence; fruit a red berry.

Moist shaded areas, often under coast pine. Alaska to California.

Iridaceae (Iris family)
This is a large family with nearly every genus containing species of value as ornamentals, most notable are the thousands of horticultural varieties of iris and gladiolus. The family can be recognized by the parallel-veined leaves which partly enfold the stem; the perianth of 6 petal-like segments; 3 stamens; inferior ovary and the fruit a capsule.

23. *Sisyrinchium californicum* (Ker.) Dryand.
Golden-eyed grass
Perennial; stems stout, broadly winged, 3 to 20 inches tall; leaves mostly basal; flowers yellow, about $1/2$ inch long.

Deflation plains. Oregon to California.

A closely related species with bluish-purple flowers, *Sisyrinchium bellum* Wats. (western blue-eyed grass) is also found on the deflation plains.

Orchidaceae (Orchid family)

This is a very large family of which most species are tropical. Economically the family is of greatest importance for its numerous ornamentals. The leaves are parallel-veined; the flowers are irregular with 6 perianth segments; the ovary inferior; stamens 1 or 2; the fruit is a capsule with abundant and minute seeds.

24. *Piperia elegans* (Lindl.) Rydb.

Dense-flowered rein orchid
Stems stout, up to 1 foot tall; stems leafy, but the leaves much reduced; inflorescence dense; flowers greenish-white, spurred.

Wet deflation plains. Alaska to California.

25. *Spiranthes romanzoffiana* Cham.

Twisted orchid or hooded ladies' tresses
Roots tuberous; stems 4-20 inches tall; leaves reduced upwards; flowers white, in a dense, twisted inflorescence.

Wet deflation plains. Alaska to southern California and on the Atlantic coast.

Salicaceae (Willow family)

This is a family of trees and shrubs of almost world wide distribution. Economically it is of slight importance as a source of paper pulp, and for its many ornamental species of willows and poplars. Members of the family have simple, alternate, deciduous leaves; unisexual flowers in catkins; no perianth; and the fruit is a capsule.

26. *Salix hookeriana* Barr.
Coast willow
Large shrub or small tree, 6 to 27 feet high; leaves dark green above, densely woolly on the under surface, at least when young, stipules small, soon failing off; flowers in catkins, appearing before the leaves.

Wet deflation plains, often with wax myrtle and coast pine in dense thickets. British Columbia to California.

There are a large number of species of *Salix* (willow); they are extremely variable and therefore difficult to separate. The several other species that occur along the coast will not be described here.

Myricaceae (Sweet Gale family)

A small family of about forty species, economically of slight importance for the aromatic wax used in making bayberry candies. Members of the family are easily recognized by the aromatic fragrance of the foliage when crushed, by yellow glandular dots on the leaves, and by the waxy-coated 1-seeded fruits.

27. *Myrica californica* C. & S.
Western wax myrtle
Shrub or small tree up to 30 feet in height; leaves evergreen, leathery, 2-5 inches long; fruit reddish-brown, approximately $1/4$ inch in diameter.

Forms dense thickets on wet deflation plains, associated with willows. Washington to northern California.

Myrica gale L. (sweet gale) is much less common, but can be found in sphagnum bogs from Alaska to Lane County, Oregon. It is a shrub up to 6 feet in height; the leaves are deciduous and the fruit is approximately $1/8$ inch in diameter.

Polygonaceae (Buckwheat family)

This family is not of major economic importance, but does include such food plants as buckwheat and rhubarb, as well as a few ornamentals and numerous weedy species. The Polygonaceae are mostly herbaceous plants with simple leaves, these usually with papery stipules forming sheaths on the stem; perianth parts 3-9, usually 6; fruit an achene.

28. *Polygonum paronychia* C. & S.
Beach knotweed
Perennial; stems somewhat woody, much-branched, trailing, 1-4 feet long; leaves narrow, $1^1/_4$ inch long or less; flowers small (about $^1/_4$ inch long), pink, usually with a green vein on each segment.

Dry deflation plains, areas of slight sand deposition. British Columbia to California.

29. *Rumex maritimus,* L.
Golden dock or seaside dock
Stems 8-20 inches long, prostrate; leaves oblong to lance-shaped, the margins crisped; flowers in crowded whorls.

Beaches and moist saline areas. Alaska to California.

Nyctaginaceae (Four O'Clock family)

This family is of only slight importance economically for a few ornamentals including the four-o'clocks and bougainvillea. Fleshy herbs with simple opposite leaves; flowers in arillary or terminal clusters; petals absent but sepals very petal-like.

30. Abronia latifolia Esch.
Yellow abronia
Perennials with stout vertical taproots; stems creeping, much-branched, leaves opposite, entire, glandular on the under surface; flowers in dense showy clusters, individual flowers small, bright yellow; fruit broadly winged.

Beaches and dunes; frequently found on large sand mounds. British Columbia to California. The large roots were used for food by the coast Indians.

A less common species is *Abronia umbellata* Lam. (pink sandverbena) which is very similar to *A. latifolia* except for the reddish-purple flower color.

Caryophyllaceae (Pink family)

This is a large family containing a number of popular ornamentals, including carnations, campions, and catch-flies. The distinguishing characteristics of the family include the opposite leaves, swollen nodes of the stem; flowers with five separate petals; superior ovary and the fruit a capsule or a utricle.

31. *Cardionema ramosissima* (Weinm.) Nels. & Macbr.

Sand mat

Mat-forming perennial; stems prostrate; leaves opposite, less than $1/2$ inch long, spine-tipped, stipules dry; flowers approximately $1/8$ inch across, borne in axillary clusters.

Beaches and dunes. Washington to South America.

32. *Cerastium arvense* L.

Field chickweed

Stems slender, often decumbent, 3-20 inches tall; leaves pubescent and glandular, narrowly lance-shaped; flowers white, the petals deeply cleft.

Beaches and cliffs. Alaska to California.

Ranunculaceae (Buttercup family)

A large family with numerous ornamentals including anemones, buttercups, columbines, larkspurs and peonies. Important characteristics of the family include numerous spirally arranged stamens; usually many simple pistils and sepals often petal-like.

33. *Ranunculus flammula* L.

Small creeping buttercup
Perennial; stems creeping; leaves alternate, simple, entire, often clustered at the rooting nodes, 1-2 inches long; flowers small, yellow.

Very wet deflation plains. Alaska to California and eastward to Labrador and Pennsylvania.

Cruciferae (Mustard family)

A large family of considerable economic importance, the numerous food plants include broccoli, Brussel sprouts, cabbage, cauliflower, radishes, rape seed, rutabagas and turnips; among the ornamentals are candy tuft, honesty, rock cress, stocks, sweet alyssum and wall flower. The family characteristics include 4 sepals, 4 petals, usually 6 stamens (4 long and 2 short) and a superior ovary.

34. *Cakile edentula* (Bigel.) Hook.

American sea rocket
Annual; herbage succulent; stems much-branched from the base, spreading or ascending; leaves alternate, obovate to oblong, coarsely

and irregularly toothed, $^3/_4$ to $1^1/_2$ inches long; flowers purple to white; fruit fleshy, 2-jointed, the upper section larger than the lower.

Beaches and adjoining dunes. British Columbia to California.

A similar species, *Cakile maritima* Scop. (sea rocket), is also found on the dunes, although it is less common. The leaves of *C. maritima* are pinnately compound with a large terminal leaflet and a few smaller remote lateral ones, or some of the leaves merely pinnately cleft.

Sarraceniaceae (Pitcher Plant family)

A small family of insectivorous bog plants. Some species are occasionally sold as novelties. Leaves pitcher-like or tubular, enlarged above; flowers long-stalked, sepals and petals 5; stamens numerous. The pitcher-shaped leaves are partially filled with an enzymatic liquid which aids in the disintegration of small organisms trapped within.

35. *Darlingtonia californica* Torr.

Pitcher plant

Perennial. Leaves hollow, tubular, enlarged and hood-like above, 1-3 feet tall; flowers solitary and nodding; petals purple; sepals larger than the petals and yellowish.

Sphagnum bogs. Lane County, Oregon, to California.

Droseraceae (Sundew family)

A small family of insectivorous plants of little economic importance, some species are sold as a novelty; most popular is the Venus-fly trap. Small perennial herbs with basal leaves; flowers with 4-5 sepals and 5 petals; fruit a many-seeded capsule. Both surfaces of the leaves are usually covered with stalked glands which aid in trapping small insects.

36. *Drosera rotundifolia L.*
Round-leaved sundew
Perennial; leaves mainly basal, reddish, leaf blades bearing gland-tipped hairs and secreting a sticky fluid by which insects are intrapped; flowers white to pinkish, petals $1/8$ to $1/4$ inch long.

Not common. Wet places on the deflation plains. Alaska to California and from the Great Lakes to the Atlantic coast, also in Eurasia.

Rosaceae (Rose family)

A large and economically important family, principally because of its fruit producing members, including apple, pear, cherry, prune, peach, apricot, blackberry and strawberry. There are also numerous ornamental species, among which are the roses, cotoneasters, firethorns, flowering cherries and hawthorns. Characteristics of the family include the alternate leaves, usually with stipules; regular flowers and numerous stamens.

37. *Fragaria chiloensis* (L.) Duch.
Coast strawberry or beach strawberry
Perennial with long stolons; leaves of 3 leaflets, leathery, shiny above, silky beneath; petals white to pinkish, $1/4$ to $1/2$ inch long; fruit red, up to $1^1/4$ inches in diameter. Fruit though small, sweet and edible.

Stable dune hummocks, dry deflation plains. Alaska to California; South America and Hawaii.

38. *Potentilla anserina* L. subsp. *pacifica* (How.) Rousi
Pacific silver weed
Low stoloniferous perennial; leaves pinnately compound, $3^1/2$ to 8 inches long with 9-31 leaflets, white silky on the under surface; flowers bright yellow, $3/4$ to $1^1/2$ inches broad.

Wet places on deflation plains often associated with slough sedge. Alaska to California.

Another species of *Potentilla* found in wet areas of the deflation plains is *P. palustris* Scop. (marsh cinquefoil). It has 3-7 leaflets and purplish-red flowers.

39. *Rubus spectabilis* Pursh
Salmon berry
Shrub 3 to 9 feet tall; stems stout, prickly; leaflets 3, 2 to 4 inches long (rarely longer); flowers red to purplish-red; fruit orange to red, raspberry-like.

Wooded areas. Alaska to California. The young, tender shoots, as well as the berries, were eaten by the Indians.

Leguminosae (Pea family)
This is probably the second largest family of flowering plants and an important source of food, including such things as peas, beans, soybeans and peanuts and such forage crops as clover, alfalfa and vetch. There are also many ornamentals in the family including acacias, locust, lupines, redbud, sweet peas and wisteria. The family (in our area) has distinctive pea-shaped flowers; alternate compound leaves, mostly with stipules; usually 10 stamens and a simple pistil.

40. *Cytisus scoparius* (L.) Link
Scotch broom
Green-stemmed shrub to 10 feet in height; leaves small, mainly of 3 leaflets; flowers yellow, in clusters of 2 or 3, $^2/_3$ to 1 inch long.

Used in dune stabilization plantings, now widely escaped and well established inland. Washington to California; naturalized from Europe.

A closely related plant is *Genista monspessulana* (L.) L. Johnson (French broom) which is very similar to *C. scoparius* except the flowers are in clusters of 3-9 and the petals are less than $1/2$ inch long. This species is much less common than *C. scoparius*.

41. *Lathyrus japonicus* Willd. var. *glaber* (Ser.) Fern.
Beach pea
Perennial; stems trailing, 1 to $3^1/2$ feet long; leaves bright green, of 3-5 pairs of leaflets, and with large stipules; flowers $3/4$ to $1^1/2$ inch long, violet.

Sandy beaches. Alaska to northern California.

42. *Lathyrus littoralis* (Nutt. ex Torr. & Gray) Endl.
Gray beach pea
Perennial; stems prostrate, $1/3$ to 2 feet long, stems and leaves white silky; leaves with large stipules, leaflets 2-8; flowers white to pink or purple, $1/2$ to $3/4$ inch long.

Beaches and areas of sand deposition. British Columbia to California.

43. *Lupinus arboreus* Sims.
Tree lupine
Stems shrubby, often 5 feet tall (rarely 10 feet); leaflets 7-11, $3/4$ to $1^1/2$ inches long, palmately arranged; flowers usually yellow, rarely white, bluish or purplish tinged.

Native of California but introduced along the Oregon and Washington coast as a sand binder and now well established.

44. *Lupinus littoralis* Dougl.
Seashore lupine
Much-branched perennial, stems prostrate, 8-24 inches long, herbage with silvery-silky hairs; leaflets 5-8; inflorescence loosely flowered, flowers purplish-blue, about $1/2$ inch long; pods 1 to $1^3/4$ inches long.

Beaches and dunes; Washington to California.

This species was called "Chinook licorice" by the early settlers, as the Indians ate the roots in times of famine.

45. *Trifolium wormskjoldii* Lehm.
Spring-bank clover
Perennial. Stems erect or prostrate, often from creeping rootstocks; leaflets 3; heads large, showy, flowers purplish-red, often white-tipped.

Dunes and deflation plains and inland. British Columbia to California.

Another species of clover, *Trifolium fucatum* Lindl. (sour clover) is an annual with bright yellow flowers which turn rose-colored and become strongly inflated in fruit. It occurs along the coast from Lane County, Oregon, to California.

46. *Ulex europaea* L.
Gorse
Shrub up to 6 feet tall, densely branched and bearing numerous stiff spines; flowers yellow, $1/2$ to $3/4$ inch long.

On hills north of Florence (near Lily Lake) and around Bandon, often forming dense impenetrable stands. Introduced from Europe and now well established.

47. *Vicia gigantea* Hook.
Giant vetch

Annual; stems 1-3 feet tall, usually climbing; leaflets 16-30, tendrils fairly well developed; flowers tan to yellowish, often tinged with purple; pods about 1 inch long or less, several-seeded.

Along the beach, often climbing in drift wood, also inland. Alaska to California. The seeds were reportedly eaten by the Indians.

Rhamnaceae (Buckthorn family)

Small trees and shrubs of slight economic importance; Cascara is obtained from the bark of *Rhamnus purshiana,* and several species are cultivated, including the buckthorns. Leaves alternate; petals, if present, curved at the tip to form a hood; stamens opposite the petals.

48. *Ceanothus thyrsiflorus* Esch.
Blue blossom

Shrub or small tree up to 24 feet tall, usually much lower; branchlets green; leaves strongly 3-veined, $3/4$ to 2 inches long; flowers small, bright blue, in large showy clusters.

Bluffs. Coos County, Oregon, to California.

Hypericaceae (St. John's Wort family)

This is a small family of only minor economic importance; it includes numerous weedy species and some cultivated species of St. John's wort. Members of the family have opposite entire leaves dotted with translucent glands; 4-5 sepals and petals, numerous stamens, usually in 3 to 5 clusters and the fruit is a capsule.

49. *Hypericum anagalloides* C. & S.

Bog St. John's wort
Stems slender, 2-10 inches tall, erect or more often creeping; leaves opposite, ovate, $1/4$ to $1/2$ inch long; flowers yellowish-orange, about $1/3$ inch across.

Bogs. British Columbia to California.

Onagraceae (Evening Primrose family)

This family is not of major economic importance, although some species are ornamentals, the most notable being the fuchsias. Herbs; flower parts in 2's or 4's; ovary inferior.

50. *Camissonia cheiranthifolia* (Sprengel) Raim

Beach evening primrose
Stems stout, prostrate; leaves grayish pubescent; flowers yellow, turning red with age, petals $1/2$ to $3/4$ inch long; fruit a four-angled capsule, becoming spirally coiled.

Beaches and dunes. Coos County, Oregon, to California.

51. *Epilobium ciliatum* Raf. subsp. *watsonii* (Barb.) Hoch & Raven

Pacific willow-herb

Perennial; stems stout, $^1/_2$ to $2^1/_2$ feet tall; leaves opposite, sharply and minutely toothed; flowers small, petals $^1/_4$ to $^1/_2$ inch long, purplish-red, fruit a capsule; seeds each bearing a tuft of hairs.

Wet deflation plains. Oregon to California.

Umbelliferae (Parsley family)

A large family of economic importance for its food products (carrots, celery, parsley and parsnips), condiments (anise, caraway and dill) and ornamentals (angelica, lace flower and sea holly). The family is distinguished by its aromatic herbage, the sheathing petioles and the typically umbellate inflorescence; flower parts in 5's; ovary inferior.

52. *Angelica hendersonii* C. & R.

Sea-coast angelica

Perennial; stems 2-6 feet tall; leaves divided into 3 or 5 principal segments, these again divided; leaves green above, white-woolly beneath, 4-10 inches long; flowers small, umbellately arranged, petals white; fruit with lateral wings.

On dune bluffs. Southern Washington to California.

A closely related species is *Angelica lucida* L. (sea watch). It also occurs on high banks and bluffs above the sea and differs from *A. hendersonii* in its glabrous leaves and in the fruit in which all the ribs are corky winged.

53. *Glehnia littoralis* (Gray) Miq. subsp. *leiocarpa* (Math.) Hult.

Beach silver-top

Perennial; plants prostrate or spreading; stems stout, short, often completely covered by sand; leaves compound, leathery, densely white-woolly on the under surface; flowers small, numerous white; fruit with broad corky wings.

Beaches and dunes in areas of sand deposition. Alaska to California.

54. *Sanicula arctopoides* H. & A.

Beach snake-root

Perennial. Plants with a deep stout taproot; herbage yellowish-green; leaves mainly basal, deeply 3-parted; flowers yellow, inconspicuous in compact head-like umbels.

Beaches and sandy bluffs. Lincoln County, Oregon, to California.

Garryaceae (Silk Tassel family)

A small family of only one genus and fifteen species; a few of the species are occasionally cultivated. Small trees or shrubs with opposite leaves; flowers small, unisexual, in dense drooping spikes.

55. *Garrya elliptica* Dougl.
Silk tassel bush
Large shrub or small tree, 6-20 feet tall; bark rough, dark; leaves evergreen, opposite, leathery, dark green above, densely hairy beneath; flowers in silky spikes up to 6 inches long; fruit deep purplish, berry-like, the outer layer pulpy at first, but becoming dry and brittle.

Bluffs and high stable dunes. Cape Perpetua to California.

Ericaceae (Heath family)

A large family important for such items as blueberries, cranberries and huckleberries, and for numerous ornamentals (azaleas, rhododendrons, heaths, and heathers). Because the family is large and diverse it is difficult to characterize. It includes saprophytic genera, herbs, shrubs and trees; the flowers are regular or nearly so and 4-5 parted; stamens 4 to 10 and the anthers often awned and often opening by pores.

56. *Arctostaphylos columbiana* Piper
Hairy manzanita or bristly manzanita
Shrub 2-9 feet tall; bark dark red, smooth; branchlets with numerous bristly hairs; leaves gray-green, $3/4$ to 2 inches long; flowers white to pale pink; fruit reddish-orange.

On dry dune areas along the edge of forests or in open pine forests. British Columbia to California.

The fruit, though dry and rather tasteless, was eaten by the Indians, both raw and cooked.

57. *Arctostaphylos uva-ursi* (L.) Spreng.
Kinnikinnick or bearberry
Low shrub; branches prostrate, forming dense mats, seldom over 6 inches tall, but several feet long; bark reddish-brown; leaves leathery, $1/2$ to 1 inch long; flowers urn-shaped, white to pink; fruit bright red.

On dunes, mostly in the open or in partial shade. Alaska to California.

The fruit is dry and tasteless when raw, but fairly palatable when cooked, the Indians used the ground-up leaves in their tobacco.

58. *Gaultheria shallon* Pursh
Salal

Spreading evergreen shrub, 1-7 feet tall, leaves leathery, 1-4 inches long, minutely toothed, dark green and shining above, paler beneath; flowers white to pink, urn-shaped; fruit dark purple to black.

Mostly in semi-shady areas, edges of dune forests. British Columbia to California.

The fruit is edible and was commonly used by the Indians who made it into syrup or dried it in cakes.

59. *Ledum glandulosum* Nutt. var. *columbianum* (Piper) Hitchc.
Pacific Labrador tea

Low shrub up to 4 feet tall; leaves alternate, lance-shaped, entire, the margins strongly rolled under; flowers in large terminal clusters, petals white.

Swampy areas. Washington to California.

The leaves of this and the following species of *Ledum* were used for making tea by the Indians and the early white settlers. However, these species are reportedly poisonous and care should be taken in their use.

Also in wet boggy areas is *Ledum groenlandicum* Oeder (Labrador tea), it differs from the above species in its rusty-woolly under surface of the leaves, while the under surface of the leaves of *L. glandulosum* is greenish or whitish and not woolly.

60. *Rhododendron macrophyllum* G. Don
Western rhododendron
Evergreen shrub up to 12 feet tall (rarely taller); leaves leathery, dark green above, paler beneath, entire, 2-8 inches long; flowers $1^1/4$ to $2^1/2$ inches wide, pink to deep rose-colored (rarely white).

In open coast pine or coast pine and Douglas-fir forests. British Columbia to California.

Another species of Rhododendron, *R. occidentale (T. & G.)* Gray (western azalea) occurs along the southern Oregon coast to California. This is a deciduous shrub the flowers of which are white with yellow stripes, often tinged with pink.

61. *Vaccinium ovatum* Pursh
Evergreen huckleberry or shot huckleberry
Evergreen shrub up to 8 feet tall; leaves leathery, toothed, $^3/4$ to $1^1/2$ inch long; flowers white to pink, $^1/8$ to $^1/2$ inch long; berry black.

On dune ridges in coast pine and Douglas-fir forests. British Columbia to California.

The berries are sweet and edible and were used in large quantities by the Indians.

Plumbaginaceae (Leadwort family)

A small family of only slight importance for ornamentals (statice, thrift, and leadwort). Herbs with basal leaves; flower parts in 5's; stamens opposite the petals; styles 5.

62. *Armeria maritima* (Mill.) Willd.
Thrift

Perennial herbs; leaves in a dense basal cluster, somewhat fleshy, linear; flowers in a dense terminal head on a naked stalk, bright pink.

Dunes and sandy bluffs. Alaska to California

Primulaceae (Primrose family)

A widely distributed family of little economic importance except for a few ornamentals, including primroses and cyclamens. Herbs; leaves usually opposite, whorled or all basal; petals fused, stamens as many as the corolla lobes and opposite them.

63. *Glaux maritima* L.
Sea milkwort

Small perennials with fleshy opposite leaves (or the upper leaves alternate); flowers in the leaf axils; corolla white or pinkish.

Salt marshes. Alaska to California and on the Atlantic coast.

Gentianaceae (Gentian family)

Economically the family is important only for its ornamentals including gentians and centauries. Herbs with opposite leaves; calyx and corolla lobes 4 or 5; ovary superior; fruit a capsule.

64. *Centaurium erythraea Ra*f.
Centaury

Annual; stems erect, 8-16 inches long, basal leaves tufted, $3/4$ to $1^1/2$ inch long, stem leaves opposite, reduced upwards; flowers many, rose-colored, $1/4$ to $3/4$ inch long.

Deflation plains. Washington to California, also in Europe.

65. *Gentiana sceptrum* Griseb.
King's gentian or staff gentian

Perennial; stems 8-28 inches tall; leaves opposite $3/4$ to $2^1/2$ inches long; flowers funnelform, 1 to $1^1/2$ inches long, blue to purplish-blue, usually with green dots within.

Very wet deflation plains with slough sedge. British Columbia to California.

A similar species, *Gentiana menziesii* Griseb. (spreading gentian), occurs on the southern Oregon coast; in this species the flowers are green or brownish outside and blue with green dots within.

Convolvulaceae (Morning-glory family)

This family includes a number of weedy species plus several ornamentals, most notable of which are the morning-glories and the wood rose. The family is also important for the sweet potato. Characteristics of the family include the usually milky sap and a plaited corolla.

66. *Calystegia soldanella* (L.) R. Br.

(=*Convolvulus soldanella* L.)
Beach morning-glory
Perennial; stems $1/2$ to 2 feet long, trailing; leaves fleshy, much broader than long; flowers trumpet-shaped, light pink to rose-colored, approximately 2 inches across; fruit a capsule.

Beaches, dunes and areas of sand movement. Washington to California, in South America and the Old World.

Polemoniaceae (Phlox family)

Members of this family are found chiefly in the western United States. It is of only slight importance economically for its ornamentals which include phlox, Jacob's ladder, gilia, and moss-pink. Petals united, stamens borne on the corolla tube at different levels; ovary superior, styles typically 3-cleft.

67. *Gilia capitata* Dougl.
Bluefield Gilia
Stems up to 1 foot tall,
glandular pubescent; leaves
much divided; flowers in
globose head-like clusters,
deep blue.
　　Dunes and bluffs. Curry
County, Oregon, to California.

Hydrophyllaceae (Waterleaf family)

A widely distributed family, but especially abundant in western
North America. A few members of the family are occasionally
cultivated including baby blue-eyes and phacelia. Petals united,
stamens borne on the corolla tube; ovary superior.

**68. *Phacelia argentea* Nels.
& Macbr.**
Silvery phacelia
Stems stout, much
branched from the base,
coarsely pubescent; leaves
silvery pubescent;
inflorescence dense, flowers
white.
　　Beaches. Curry County,
Oregon, to California.

69. *Phacelia bolanderi* Gray

Bolander's phacelia

Stout perennials from slender rhizomes; coarsely pubescent and sometimes glandular; leaves mostly compound, with a terminal lobe and a smaller, lower pair; flowers bright blue, $^1/_2$ to $^3/_4$ inch long.

Beaches and bluffs, often in moist areas. Southern Oregon to California.

Scrophulariaceae (Figwort family)

A large family of little economic importance except for the drug plant *Digitalis* (Foxglove) and for numerous ornamentals, the most important of which are snapdragon, speedwell, monkey flower, and foxglove. Flowers irregular, the petals fused; stamens 2, 4 or 5; ovary superior, fruit a capsule.

70. *Castilleja ambigua* Hook. & Arn.

(= *Orthocarpus castillejoides* Benth.)

Slender stems 2-12 inches tall; leaves often 2- or 6-lobed; bracts of the inflorescence yellow or white-tipped; corolla light yellow with purple spots.

Salt marshes. British Columbia to California.

One other species of Castilleja is common along the coast, *C. affinis* Hook. & Arn. subsp. *litoralis* (Penn.) Chuang & Heckard (Pacific Paintbrush). It is sometimes woody at the base, 4-24 inches tall and finely pubescent, the leaves are usually entire although occasionally with one or more short lobes; the bracts are broad and tipped with scarlet; flowers yellowish-green with red. This species occurs on bluffs and old dunes from Oregon to California.

Two other related species, now placed in the genus *Triphysaria*, occur on the southern Oregon coast: *T. erianthus* (Benth.) Chuang & Heckard is 2-16 inches tall and finely pubescent and often glandular; the leaves are broadly clasping at the base and narrowly lobed; the bracts of the inflorescence are purple-tinged and the corolla is purple, yellow, and white. *T. versicolor* Fisch. & Mey subsp. *faucibarbata* (Gray) Chuang & Heckard is similar to *T. erianthus* except the stems are glabrous and the corolla is yellow.

71. *Digitalis purpurea* L.
Foxglove
Stems up to 9 feet tall (rarely taller); leaves large, ovate, with a white soft pubescence on the under surface; flowers in a long narrow inflorescence, white or rose-purple with dark spots within.

Bluffs and roadcuts. Naturalized from Europe.

72. *Mimulus guttatus* DC. var. *grandis* Greene
Common monkey flower
Stems stout, 2-40 inches tall; leaves opposite; flowers irregular, 1-2 inches long, yellow, the throat usually red-spotted.

Wet deflation plains and wet ocean bluffs.

Washington to California. The Indians reportedly used the young leaves for salad.

One other species of *Mimulus* is abundant on the immediate coast. This is *M. dentatus* (coast monkey flower), which occurs in wet areas of the deflation plains, as well as inland, from Washington to California. It is a perennial with slender rhizomes and stolons; stems 5-24 inches tall; flowers less than 1 inch long, bright yellow, spotted red within and sometimes red-tinged outside.

Orobanchaceae (Broom-rape family)

A small family of no particular economic importance although several species are parasitic on crops and do cause some damage. Non-green root parasites: leaves reduced and scale-like; petals united, forming a 2-lipped tube: ovary superior; fruit a capsule.

73. *Boschniakia hookeri* Walp.
Small ground-cone
Whole plant reddish-brown, stems simple, 3-6 inches tall, thick and fleshy; leaves scale-like, usually overlapping; flowers sessile, concealed by the bracts. Parasitic on *Gaultheria shallon.*
　　Dune forests. British Columbia to Calilornia.

Plantaginaceae (Plantago family)

This family is not particularly important except for some noxious lawn weeds. Characteristics of the family include the rosette of basal leaves with apparent parallel venation, the spicate or capitate inflorescence on a naked stalk and the 4-lobed membranous corolla.

74. *Plantago maritima* L.
Seaside plantain

Leaves basal, linear, succulent; pubescent, 2-12 inches tall; spike dense, 1- 5 inches long; bracts fleshy. This is a variable species with two varieties in our area. Variety *californica* (Fern.) Pilg. occurs principally on cliffs and bluffs and has leaves 1-3 inches long, often toothed. Variety *juncoides* (Lam.) Hult. occurs on beaches and in salt marshes and has more elongate leaves, up to 10 inches long.

Cliffs, bluffs, beaches and salt marshes. Alaska to California.

The young leaves are edible and may be cut up and cooked like green beans or used fresh in salads.

At least three other species of *Plantago* are common along the coast: P. *subnuda* Pilger (tall coast plantain), P. *coronopus* L. (cut-leaved plantain), and P. *elongata* Pursh.

The species of *Plantago* may be separated as follows:

1a. Leaves divided into narrow segments; plants restricted to the Curry County coast ....*P. coronopus*

1b   Leaves entire or toothed, but never divided.

2a. Plants annual .... *P. elongata*

2b. Plants perennial.

3a. Leaves linear or nearly so; corolla tube pubescent ....
.... *P. maritima*

3b. Leaves broadly oblanceolate; corolla tube glabrous ....
....*P. subnuda*

Caprifoliaceae (Honeysuckle family)

This family contains numerous ornamental shrubs including honeysuckle, elderberry, laurestinus, snowball, abelia, wayfaring-tree and weigela. Distinctive features of the family include opposite leaves, united petals, inferior ovary and the fruit a berry or a drupe.

75. *Lonicera involucrata* (Rich.) Banks.

Black twinberry

Shrub $1^1/2$ to 12 feet tall; leaves more or less ovate; flowers axillary and borne in pairs in large bracts; corolla yellow, often purplish-tinged; fruit black.

Wooded areas. Alaska to California.

An introduced species, *Lonicera etrusca* Santi (Etruscan honeysuckle), has become established along the coast from Lane County southward. Its flowers are yellow, often tinged with purple and are clustered, but not borne in pairs as in *L. involucrata*.

Compositae (Sunflower family)

This is the largest family of vascular plants. It is important as a source of food, e.g. lettuce, artichoke and endive, and it includes a number of noxious weeds, plus many ornamental species, notably, asters, chrysanthemums, dahlias and marigolds. The flowers are borne in involucrate heads; the anthers are united to form a tube; the ovary is inferior and the fruit is an achene.

76. *Achillea millefolium* L.
Yarrow
Perennial; stems 1-3 feet
tall; leaves pinnately divided;
heads many, small and
clustered in a flat-topped
inflorescence; flowers white
or rarely pinkish.

Deflation plains.
Throughout the United
States and Eurasia.

77. *Ambrosia chamissonis* (*Less.*) Greene
(= *Franseria chamissonis* Less.)
Silver beach-weed
Perennial from a stout
taproot; stems prostrate or
nearly so, 2-4 feet long;
leaves silvery with silky
hairs; heads of two kinds,
the pistillate axillary, the
staminate in a terminal
raceme or occasionally the
two types intermixed;
involucral bracts of the
pistillate flowers becoming bur-like in fruit, armed with short
prickles.

Beaches and dunes and on hummocks in areas of moving
sand. British Columbia to California.

A variety of this species is similar but less silvery and with
dissected leaves.

78. *Anaphalis margaritacea* (L.) B. & H.

Pearly everlasting
Perennial; stems ³/₄ to 3 feet tall, white woolly; leaves narrow, 1-3 inches long (rarely longer), the under surface white woolly; heads small, numerous, the involucral bracts papery, white.

Dry stabilized dunes. Alaska to California, on the Atlantic coast and in Eurasia.

79. *Artemisia pycnocephala* (Less.) DC.

Beach sagewort
Perennial; stems woody at the base; leaves 2 or 3 times divided into narrow linear divisions; herbage densely covered with long silky-woolly hairs; inflorescence dense, leafy; flowers in small heads.

Beaches. Southern Oregon to California.

A similar species occurs on the northern beaches; it is *Artemisia campestris* L. (silky field wormwood) and has appressed silky hairs as opposed to the silky-woolly hairs of *A. pycnocephala*.

80. *Aster chilensis* Nees
Common California aster
Stems 1 to 3 feet tall; leaves clasping the stem, entire to more or less toothed; flowers in heads, the ray flowers approximately $1/2$ inch long, purplish-blue. An extremely variable species.

Deflation plains. Oregon to California.

81. *Baccharis pilularis* DC.
Chaparral broom
Shrub 2 to 5 feet tall; leaves alternate, sparsely toothed, thick; flowers in small dense clustered heads.

Banks and cliffs. Oregon to California.

82. *Cotula coronopifolia* L.
Brass buttons
Low, much-branched perennial; stems weak, fleshy; leaves oblong in outline, clasping the stem, the lower deeply cleft, the upper often entire; heads yellow.

Beaches and bogs. Washington to California.

83. *Erechtites minima (Poir.)* DC.
Australian fireweed
Annual; stems 2-8 feet tall; leaves sharply and finely toothed, green above, white-woolly on the under surface, at least when young; flowers minute, in numerous small heads, yellow.

Moist places on dunes, especially in areas of stabilization plantings. Oregon to California; native of Australia and New Zealand.

84. *Erigeron glaucus* Kerr.
Seaside erigeron
Perennial. Stems more or less decumbent from creeping rhizomes; leaves thick, basal leaves obovate, toothed, stem leaves reduced, spatulate and entire or few-toothed; heads 1-6 per stem; rays light purple to white.

Banks and cliffs. Oregon to California.

85. *Gnaphalium purpureum* L.
Purple cudweed
Herbage densely white-woolly; leaves oblanceolate, reduced upwards; inflorescence narrow, heads small, involucral bracts brown or purplish; corollas purplish.

Deflation plains. British Columbia to California.

Another species, *Gnaphalium stramineum* Kunth (cotton-batting plant), occurs in exposed situations, on cliffs and bluffs and in the dunes. It is also densely white-woolly, but the involucral bracts are distinctly yellowish rather than brown or purplish as in *G. purpureum*.

86. *Hypochaeris radicata* L.
Gosmore or false dandelion
Leaves all basal, pinnately lobed; flowering stalk $^{1}/_{2}$ to 2 feet tall; several headed; flowers yellow.

Deflation plains. Common weed, introduced from Europe.

87. *Jaumea carnosa* (Less.) Gray
Rhizomatous perennial; stems lax; leaves opposite, united at the base, narrow, succulent; heads solitary; flowers yellow.

Tidal flats and salt marshes. Vancouver Island to California.

88. *Leontodon taraxoides* (Villars) Merat

(= *Leontodon nudicaulis* (L.) Merat)
Bristly hawkbit
Leaves basal, usually shallowly lobed, pubescence stiff; heads solitary; flowers yellow.

   Deflation plains, also inland as a weed in disturbed sites. Native of Europe, now established throughout the Pacific states.

89. *Solidago simplex* Kunth var. *spathulata* (DC.) Cronq.

Sticky goldenrod
Stems stout, $1/2$ to 2 feet tall; leaves alternate, toothed, more or less glandular; flowers yellow, in numerous small heads.

   Dry sandy areas of little or no sand movement; frequently associated with red fescue, various mosses and bearberry. Oregon to California.

90. *Tanacetum camphoratum* Less.

Seaside tansy or western tansy
Stems stout, $3/4$ to 2 feet tall; leaves finely divided, bearing scattered long white hairs; heads yellow, numerous, $1/4$ to $3/4$ inch across.

   Ocean bluffs, active dune areas, often on hummocks. Washington to California.

Glossary

Achene: A small dry 1-seeded fruit which does not split open on regular lines at maturity.

Alternate: Placed singly, said of leaves when there is only one per node.

Annual: A plant which completes its life cycle in one year, then dies.

Anther: The pollen-bearing portion of the stamen.

Awn: A bristle-like appendage.

Axillary: Borne in the angle between the leaf and a stem.

Basal: Situated at the base; said of leaves arising from the base of the stem.

Berry: A single (not aggregate) fleshy fruit containing one or usually many seeds, as a grape or a tomato.

Blade: The expanded portion of a leaf or a petal.

Bract: A leaflike or scalelike structure usually in the inflorescence; a modified leaf subtending a flower.

Capitate: Head-like; in dense clusters.

Capsule: A dry fruit from a compound pistil which splits open at maturity.

Carnivorous: Applied to plants able to trap insects and other small organisms.

Catkin: A scaly spike of unisexual flowers; an ament; e.g. Willow.

Channeled: Grooved.

Clasping: Partly or completely surrounding the stem.

Cleft: Divided to about the middle.

Collar: The area on the outer side of the leaf where the sheath and blade join (Gramineae, Cyperaceae and Juncaceae).

Compound: Composed of two or more similar but separate parts, said of a leaf when it is divided into leaflets, of an inflorescence when it is branched, or a pistil when it has more than one chamber, style or placenta.

Cone: Seed or pollen bearing structure of most gymnosperms.

Corolla: The conspicuous part of the flower, usually showy and colored. The petals collectively.

Crisped: Wavy or curled.

Deciduous: Not persistent or evergreen, but failing away at maturity; said of plants whose leaves drop off in the fall.

Decumbent: Lying on the ground at the base but then turning upward

Dissected: Deeply cut or divided into numerous segments.

Distinct: Separate, not fused or united.

Drupe: A fleshy 1-seeded fruit with the seed enclosed in a stony covering, the "pit," e.g., prune, olive.

Entire: Smooth and uninterrupted; without lobes or teeth.

Evergreen: Green throughout the year.

Fertile: Capable of producing seeds or spores.

Fibrous roots: A root system consisting of numerous slender, frequently branched roots.

Filiform: Thread-like; long and slender.

Floret: Individual flower, especially in Gramineae and Compositae.

Fruit: The matured ovary containing the seed, plus any other parts developing with it.

Funnelform: Gradually widened upward; shaped like a funnel.

Glabrous: Not bearing hairs.

Gland: Usually a small secreting body; glands may be embedded or on the surface or at the tip of hairs.

Glumes: The pair of chaffy bracts at the base of the spikelet in the Gramineae.

Head: A compact cluster of flowers; technically all the flowers are sessile and the outer flowers bloom first.

Herb: Herbaceous: a plant lacking perennial woody tissue above ground, i.e., the stems above ground die back each year.

Herbage: Vegetative parts of the plant collectively.

Inferior: Below, said of an ovary fused with the bases of the sepals and petals, seemingly borne below the sepals.

Inflorescence: The flower cluster or the arrangement of the flowers.

Insectivorous: Capable of trapping insects.

Involucre: The whorl of bracts at the base of the flower cluster.

Irregular: Made up of parts that are unequal in size or shape.

Keel: A ridge.

Lateral: Relating to the side; coming from or situated on the side.

Lax: Weak or loose.

Leaflet: One of the leaf-like divisions of a compound leaf.

Lemma: In the Gramineae the lower of the two bracts immediately surrounding the stamens and pistil.

Ligule: In the Gramineae an appendage on the inner side of the leaf where the sheath and blade join.

Linear: Narrow and flat with the sides parallel.

Lip: One of the parts of a 2-lipped flower.

Lobe: A shallow, more or less rounded division; leaves, petals, etc., are said to be lobed.

Membranous: Thin and more or less transparent.

Midrib: The main vein of a leaf blade.

Minute: Very small.

Node: A joint; a place on the stem which normally bears leaves.

Nutlet: A small nutlike fruit.

Oblanceolate: Broadest at the upper end and elongating and tapering to the base.

Oblong: Longer than broad and with the sides nearly parallel.

Obovate: Broadest at the upper end and narrowing abruptly to the base.

Opposite: Arranged in pairs; e.g., leaves which are two at a node.

Orbicular: Circular in outline.

Ovary: The ovule or seed bearing part of the pistil.

Ovate: Egg-shaped, the broader end at the base.

Palea: In the Gramineae, the upper of the two bracts immediately surrounding the stamens and pistil.

Palmate: Spreading like the fingers from the palm: in venation, all veins coming from one basal point; in compound leaves, the leaflets all coming from one point.

Parasitic: Deriving food from another living organism.

Perennial: Said of herbaceous plants that live for three or more years; usually does not flower the first year.

Perianth: The sepals and petals collectively, particularly if they are similar.

Petal: One unit of the corolla, usually colored and more or less showy.

Petiole: The stalk of the leaf.

Pinnate: With parts arranged along the sides of a central axis; as in a compound leaf with leaflets on the sides of the main axis.

Pistil: The central organ of the flower, composed of ovary, style and stigma.

Pith: The soft spongy center of most stems.

Prickle: A sharp weak thorn-like outgrowth on the stem or fruit.

Prostrate: Lying on the surface of the ground; usually applied to stems.

Pubescent: Having hairs.

Raceme: An inflorescence in which the flowers are borne on individual stalks of nearly equal length; the lower flowers bloom before the terminal ones.

Rank: Arranged in vertical rows; with leaves, the arrangement on a stem in relation to one another.

Ray flowers: The outer flowers of the head in Compositae, usually with strap-shaped corollas.

Reduced: Diminished in size.

Regular: Said of flowers, when all the petals or all the sepals are of equal size and shape.

Rhizome: An underground stem typically elongated, functioning in food storage and vegetative reproduction.

Scale: A small modified leaf.

Scape: A leafless stalk arising from the ground and bearing the inflorescence.

Sepal: One of the parts of the outer whorl of the flower, usually green.

Sessile: Without a stalk; as leaves without petioles or flowers without pedicels.

Sheath: Part of the leaf which enfolds the stem; usually more or less tubular in shape. Sheathing: enfolding the stem.

Simple: Undivided, unbranched, not compound.

Spike: An elongated flower cluster in which the flowers are sessile.

Spikelet: A term applied to the main floral unit of the grass and sedge families; the small spike-like unit of the inflorescence of these families.

Spine: A sharp rigid thorn-like outgrowth.

Sporangium: A case in which the spores are borne; plural: sporangia.

Spore: A generally single-celled structure capable of producing a new plant.

Spur: A hollow, sac-like projection.

Stamen: The pollen bearing organ (anther) plus its stalk (filament).

Sterile: Lacking functional sex organs or sporangia; not capable of sexual reproduction.

Stigma: The part of the pistil that receives the pollen.

Stipules: A pair of appendages at the base of a petiole.

Stolon: A stem running along the surface of the ground that roots and produces a new plant; a runner.

Subtend: Borne at the base and close to the structure in question.

Succulent: Fleshy, juicy, soft and thick in texture.

Superior: Said of an ovary that is borne above and/or free from the other floral structures.

Tendril: A slender modification of a stem or a leaf by which a plant clings or coils about a support.

Toothed: Provided with tooth-like projections, as on the margin of a leaf.

Tufted: Producing abundant, closely clustered stems.

Umbel: An inflorescence whose branches all arise from one point.

Unisexual: Of one sex, bearing only stamens or only pistils.

Venation: The arrangement of the veins.

Whorl: Three or more at a node; as in leaves or flowers.

Wing: A thin flattened extension of an appendage or an organ.

Index

Entries in bold refer to main description of species

116   PLANTS OF THE OREGON COASTAL DUNES

Carex, 21
    lenticularis, 36; var. *limnophila*, 68
    livida, 68
    macrocephala, 1, 27, 50, **67**, 68
    obnupta, 36, 50, **68**
    pansa, 68
    viridula, 35, 68
Caryophyllaceae, 77-78
Castilleja
    affinis subsp. *litoralis*, 99-100
    ambigua, 35, 55, **99**
Ceanothus thyrsiflorus, 3, 48, **87**
Cedar, 31
    giant, 62
    Port Orford, **61**
    western red, 22, 29, 30, 62
Centaurium erythraea, 35, 53, **96**
Centaury, 35, **96**
Cerastium arvense, 53, **78**
Chamaecyparis lawsoniana, 47, **61**
Chaparral broom. *See* Broom
Chickweed, field, **78**
Cinquefoil, marsh, 82
Cladonia, 30
Clover, 21
    sour, 86
    spring-bank, 35, **86**
Club-moss
    bog, 36, **59**
    family, 59
Compositae, 103-9
Conifer family, 60-62
Cotton-batting plant, 108
Cotton-grass, russet, **69**
Cotula coronopifolia, 54, 55, **106**
Convolvulaceae, 97
Convolvulus soldanella, **97**
Cruciferae, 79-80
Cudweed, purple, **107**
Cupressaceae, 61-62
Cyperaceae, 67-69
Cypress family, 61-62
Cytisus scoparius, 2, 39, 48, **83**, 84

Dandelion, false, 34, **108**
Darlingtonia californica, 53, **80**
Deer-fern, **57**
Digitalis purpurea, 55, **100**
Dock
    golden, **76**
    seaside, **76**
Douglas-fir, 22, 23, 24, 28, 29, 30, **61**
Drosera rotundifolia, 35, 54, **81**
Droseraceae, 81
Dune-grass, American, 1, 27, **66**

Eleocharis macrostachya, 50, **69**
Elymus mollis, 1, 27, 50, **66**
Epilobium
    ciliatum, 35, 53; subsp. *watsonii*, **89**
Erechtites minima, 33, 55, **107**
Ericaceae, 91-94
Erigeron glaucus, 54, **107**
Erigeron, seaside, **107**
Eriophorum chamissonis, 50, **69**
Eurhynchium oreganum, 29
Evening primrose
    beach, **88**
    family, 88-89
Everlasting, pearly, 28, 34, **105**

False Dandelion. *See* Dandelion
Fern,
    family, 57-58
    licorice, 58
    sword, 58
    western bracken, 28, **58**
    See also Deer-fern, Grape-fern
Fescue, 21
    alta, **66**
    red, 28, 30, 31, 34, 66
    tall, **66**
Festuca, 21
    arundinacea, 51, **66**
    rubra, 28, 34, 66
Figwort family, 99-101
Fireweed, Australian, 33, **107**
Four O'Clock family, 77
Foxglove, **100**